Memoirs of the American Mathematical Society

Number 222

Maurice J. Dupré

The classification and structure of C*-algebra bundles

Published by the

AMERICAN MATHEMATICAL SOCIETY

Providence, Rhode Island, USA

September 1979 · Volume 21 · Number 222 (end of volume)

ABSTRACT

A C*-bundle (p,B,X) is a type of fibre space $p : B \to X$ such that each fibre $p^{-1}(x)$ is a C*-algebra. However $p^{-1}(x)$ may vary in dimension as x varies in X, even when X is connected. We develop a "homotopy" type classification theory for C*-bundles having primarily finite dimensional fibres. A $C^*(m,n)$-bundle over the pair (X,A) is a C*-bundle (p,B,X) such that $p^{-1}(x) \cong L(\mathbb{C}^m)$ for $x \in A$ and $p^{-1}(x) \cong L(\mathbb{C}^n)$ otherwise. As a special case our theory shows that if X is a compact metric space, C^+X the upper cone of the suspension SX, then the isomorphism classes of $C^*(m,n)$-bundles over (SX,C^+X) are one one-to-one correspondence with members of $[X,C^*U(m,n)]$, where $C^*U(m,n)$ is the compact manifold of injective *-homomorphisms $L(\mathbb{C}^m) \to L(\mathbb{C}^n)$. The lower homotopy of $C^*U(m,n)$ is calculated for illustration. The results are applicable to the classification of C*-algebras.

1980 Mathematics Subject Classification. Primary: 46L05.
Secondary: 55R05, 55R10, 55R15, 55R25.

Library of Congress Cataloging in Publication Data

Dupré, Maurice J
The classification and structure of C*-algebra bundles.

(Memoirs of the American Mathematical Society ; no. 222)
Bibliography: p.
1. C*-algebras. 2. Fiber bundles (Mathematics)
I. Title. II. Series: American Mathematical Society. Memoirs ; no. 222.
QA3.A57 no. 222 [QA326] 510'.8s [512'.55] 79-17975
ISBN 0-8218-2222-5

TABLE OF CONTENTS

INTRODUCTION

In the representation theory of C*-algebras and locally compact groups, certain fibre spaces arise naturally, all of whose fibres are Banach spaces. These fibre spaces , our objects of study, we call <u>Banach bundles</u> (precise definitions will be provided in the next section) following Fell [18], although the older term "continuous field of Banach spaces" is still common in the literature. In order to motivate our study, we recall that if A is a commutative Banach algebra and $\mathrm{Max}(A)$ its space of maximal ideals with the hull-kernel topology, then much of the algebraic structure of A is reflected in topological properties of $\mathrm{Max}(A)$. The Gelfand homomorphism $A \to C_0(\mathrm{Max}(A))$ is the natural attempt to recover A from $\hat{A}$. In the case where A is a commutative C*-algebra, this attempt succeeds, as is well known, giving a *-isomorphism $A \cong C_0(\mathrm{Max}(A))$. The result is a most useful and beautiful duality between C*-algebras and locally compact Hausdorff spaces. It is only natural that such a program should be and has been attempted for non-abelian C*-algebras. To begin with, the maximal ideal space is replaced by the space $\mathrm{Prim}(A)$ of primitive ideals, for a general C*-algebra A. Of course, if A is commutative, then $\mathrm{Prim}(A) = \mathrm{Max}(A)$, and the Gelfand isomorphism can be viewed as sending each $a \in A$ to the function $\hat{a}$ with $\hat{a}(I) = a + I \in A/I \cong \mathbb{C}$, $I \in \mathrm{Max}(A)$. For the general case, our Gelfand transform $\hat{a}$ is a function on $\mathrm{Prim}(A)$ with $\hat{a}(P) = a+P \in A/P$, $P \in \mathrm{Prim}(A)$. Our immediate problem then is that the factor algebras A/P, $P \in \mathrm{Prim}(A)$ can vary as P varies. In the case where $\mathrm{Prim}(A)$ is Hausdorff, it is possible to put a topology on the disjoint union

$$E = \bigcup\{A/P : P \in \mathrm{Prim}(A)\}$$

so that we obtain a Banach bundle ξ over $\mathrm{Prim}(A)$, with A *-isomorphic to the algebra of global sections vanishing at infinity, $\Gamma_0(\xi)$.

In this case, the Banach bundle is in fact a C*-algebra bundle or C*-bundle, for short, a type of Banach bundle all of whose fibres are C*-algebras. The isomorphism $A \cong \Gamma_0(\xi)$ is of course the more general Gelfand transformation. This beautiful result was obtained in the work of Fell [19], although special cases were known earlier. More generally, the theory of [19] shows that any C*-algebra A is classifiable as a C*-algebra of sections of some C*-bundle ξ over a compact Hausdorff space $H(\mathrm{Prim}(A))$ associated to $\mathrm{Prim}(A)$. We therefore have $A \subset \Gamma(\xi)$, but some sections may not come from A via the transforms, even though the sections which do can be nicely characterized. As beautiful as it is, the theory has an obvious major setback. The resulting C*-bundles, used to represent C*-algebras in general, must be considered unfamiliar mathematical territory. Any serious mathematician must wonder: why bother representing C*-algebras as section algebras of C*-bundles when we apparently know so much more about C*-algebras than C*-bundles? If one entertains an optimistic view of this general Gelfand theory one must still ask: what are these C*-bundles - what is their structure and how can they be classified? The extent to which we can answer these questions give the limit to the applicability and usefulness of the theory.

If all its fibres are Hilbert spaces, a Banach bundle is called a Hilbert bundle. The work of [9,13,14,15] gives a substantial theory for the structure and classification of a large class of Hilbert bundles, even though interesting questions remain. Through the work of [7,9], the results of the classification of Hilbert bundles are applicable to the classification of separable continuous trace C*-algebras. However, for a large class of type I C*-algebras, the sectional representation of [19] lays bare much of the interesting structure - theoretically all of it. In this paper we shall develop a theory for classifying C*-bundles in terms of homotopy invariants.

The older term "continuous field of Banach spaces" goes back to the inventor, Godement [21] who used Hilbert bundles in the study of group representations. The reason for our terminology is mainly due to the fact that these fibre spaces can profitably be thought of as bundles, although they are more general than the ordinary fibre or vector bundles familiar from topology. We shall emphasize topological methods and constructions, emphasizing the bundle or fibre space viewpoint over the equivalent but often clumsier continuous field outlook.

If ξ is a C*-bundle over X, then the fibre $\xi(x)$ of ξ over x may vary in dimension or isomorphism type as x varies in X, even though X may be connected. This phenomena is both the main problem and the main source of interest in these bundles. Consequently, we call the <u>order</u> of ξ the cardinality of the set of all isomorphism classes of fibres of ξ. The work of [19] or [29] shows that if, for some finite n, every fibre of ξ is $M(n)$, the C*-algebra of all $n \times n$ complex matrices, then ξ is locally trivial and is therefore a fibre bundle in the ordinary sense of [28]. If ξ is first order (also called <u>homogeneous</u>) with finite dimensional fibre F, then ξ is locally trivial [19]. Classical topological methods now give a very satisfactory theory of classification for these homogeneous bundles. Now suppose that ζ is any C*-bundle over the paracompact base space X. Let F be a fibre of ζ having minimal dimension and let A be the set of points $x \in X$ with $\zeta(x) \cong F$. Suppose further that F is finite dimensional. Then $\zeta|A$ is locally trivial so we can assume its structure to be well known. Put $\xi = \zeta|X\backslash A$. Obviously the isomorphism class $[\xi]$ of ξ is an isomorphism invariant of ζ, as is $[\zeta|A]$. Thus the question arises as to whether the structure and isomorphism class of ζ can be determined by $[\zeta|A]$, $[\xi]$, and appropriate "pasting" data describing how $\zeta|A$ and ξ together form ζ. Answering this question will then classify second order bundles in terms of first order bundles as well as theoretically giving an inductive iteration proceedure for classifying finite order C*-bundles. The main aim of this paper is to give such a pasting classification for certain classes of C*-bundles having primarily finite dimensional fibres, and thus classifying the resulting second order bundles. To do this will require the development of a certain amount of machinery, which is of interest in its own right. In a subsequent paper, we shall develop a type of cohomology theory as announced in [12] for classifying certain finite order C*-bundles directly and which gives the present classification formula in the second order case. Because of this, we shall, where it is natural, develop some of the machinery here in a more general form than necessary for the proof of our main theorem, since it will be useful in later developments of the theory. Moreover, the development of our machinery will also serve to illuminate the structure and inner workings of C*-bundles. Our reason for treating the second order theory separately is two-fold. First, the second order theory applies to give a classification of arbitrary second order C*-bundles having finite dimensional fibres, a result so far not obtainable by simply specializing the cohomology theory to the second order case and more general than announced in [12]. Second, the development of the cohomology theory would require the introduction of a considerable amount of complication which is completely unnecessary in the second order case. Our second order theory can also be viewed as a theory of extensions of one C*-bundle by another with given base space X, and is thus a C*-bundle "version" of a theory for $\mathrm{Ext}_X(B,C)$ as defined in [2]. Extensions of bundles have also been studied by Delaroche [6] but from a rather different point of view requiring in applications such great restrictions on base spaces as to make them homotopically uninteresting and allowing much greater generality in fibre types. In fact there is almost no actual overlap of our work with that work.

The paper is organized in somewhat circuitous fashion. The first section is devoted to some preliminaries, and establishment of some basic notation and terminology. In the first section we state our main theorem. In the second section we prove a general result concerning extensions of sections. We show that an arbitrary bundle over a paracompact Hausdorff space having locally a property which we call the NEP (neighborhood extension property) necessarily has the NEP globally. In section three we show that certain bundles arising as bundles of C*-bundle maps have the NEP. This can be viewed as a generalization of Fell's result [19] that any homogeneous C*-bundle with finite dimensional fibre is locally trivial. In the fourth section we develop a theory of tensor products of C*-bundles. In the fifth section we discuss commutants of C*-subbundles and obtain a type of tensor product decomposition. Section six contains the details of the pasting construction and section seven contains theorems showing when homotopically equivalent C*-bundles are isomorphic. Finally, section eight contains the proof of the main theorem. In section nine we give some applications and computations related to the main theorem, and the final section gives applications to stable isomorphisms of C*-bundles, in analogy with the concept of stable isomorphism for C*-algebras introduced in [3], in the Brown, Douglas, Fillmore theory of Ext as in [4].

ACKNOWLEDGEMENT

I would like to take this opportunity to express my gratitude for the support of the National Science Foundation. I would also like to thank Larry Brown, Ronald Fintushel, Karl Hofmann, Terry Lawson, Javier Thayer, and the members of the Tulane topological algebra seminar for much useful and stimulating conversation and correspondence. I also thank the referee for pointing out numerous minor errors. Any that remain are the sole responsibility of the author. Finally, I am greatly indebted to Meredith Mickel for her painstaking efforts in typing the manuscript.

Maurice J. Dupré
January, 1977
Department of Mathematics
Tulane University
New Orleans, Louisiana

THE CLASSIFICATION AND STRUCTURE OF C*-ALGEBRA BUNDLES

1. THE MAIN THEOREM

Before we can state our main results precisely, we must establish some notation, make some definitions, and recall some relevant facts.

By a *bundle* or *fibre space* we mean simply a triple $\xi = (p,B,X)$ where $p : B \to X$ is a continuous map, whereas we use the term *fibre bundle* in the sense of [28] or [23]. If ξ is a fibre space over X, say $\xi = (p,B,X)$, then we denote by $\xi(x) = p^{-1}(x)$, the fibre of ξ over $x \in X$. As usual, $\Gamma(\xi)$ denotes the set of continuous cross-sections of the bundle ξ. The members of $\Gamma(\xi)$ will henceforth be called simply *sections*. We will use primarily the notation and terminology of [13,14,15], but we use $|A$ to denote restriction to $A \subseteq X$. It will also often be convenient to set $\Gamma(A,\xi) = \Gamma(\xi|A)$ when ξ is a bundle over X, and $A \subseteq X$.

A *Banach bundle* [18] is a bundle $\xi = (p,B,X)$ where $p : B \to X$ is a continuous open surjection with each fibre $\xi(x)$, $x \in X$, the underlying set of a Banach space, and such that

(1) the maps $B \oplus B \to B$, $\mathbb{C} \times B \to B$, $B \to \mathbb{R}$ given by addition, scalar multiplication, and norm, respectively, in each fibre, are continuous, and

(2) if (b_α) is a net in B with $p(b_\alpha) \to x \in X$ and $||b_\alpha|| \to 0$, then $b_\alpha \to 0(x)$, the zero in $\xi(x)$. Here $B \oplus B = (p \times p)^{-1}(\Delta_X)$, where Δ_X is the diagonal in $X \times X$. We call ξ a *C*-bundle* [15] if in addition each fibre $\xi(x)$ is a C*-algebra and

(3) the maps $B \oplus B \to B$, $B \to B$ given by multiplication and *-operation, respectively, in each fibre, are continuous.

If ξ is a Banach bundle, then $\Gamma(\xi)$ is a vector space and $\Gamma_b(\xi)$, the vector subspace of bounded continuous sections of ξ, is a Banach space. Put $\Gamma_b(\xi|A) = \Gamma_b(A,\xi)$ for $A \subset X$. If ξ is a C*-bundle, then $\Gamma_b(A,\xi)$ is a C*-algebra. A *Hilbert bundle* is a Banach bundle each of whose fibres is a Hilbert space. For the elementary

---

Received by the editor May 10, 1977.

properties of Banach bundles, Hilbert bundles, and C*-bundles, we refer the reader to [9,18,13,14,15].

We refer the reader to the introduction of [14] for a general construction for producing many examples of Banach bundles and C*-bundles with varying fibres.

We say that a C*-algebra is a *simple dual* or *elementary* C*-algebra if it is isomorphic to the C*-algebra of compact operators on some Hilbert space. We say ξ is a *simple* C*-bundle, or SC*-*bundle* for short, if ξ is a C*-bundle with all fibres simple dual. We say ξ is a C_1^*-bundle if each fibre has an identity and if the identity section is continuous. Thus an SC_1^*-bundle is a simple C*-bundle with continuous identity field. On the other hand, we shall say ξ is an SC_0^*-bundle to mean ξ is a simple C*-bundle satisfying Fell's condition [19] or [9], namely that the bundle admits local rank one projection fields about each point of the base space. The notions of Banach bundle map and C*-bundle map are defined in the obvious way [13], [15]. A Banach bundle (C*-bundle) map is called *strong* if it is isometric and bijective on fibres. Thus, if $\phi : \xi \to \zeta$ is a C*-bundle map then it is strong if it is bijective on fibres. $\phi : \xi \to \zeta$ is a C_1^*-bundle map if ξ and ζ are C_1^*-bundles and ϕ is identity preserving on fibres. $\phi : \xi \to \zeta$ is a C_0^*-bundle map if ξ and ζ are SC_0^*-bundles and ϕ is trace preserving on fibres.

By the *order* of a Banach bundle (C*-bundle) we mean the cardinality of the set of isometric isomorphism (*-isomorphism) classes of its fibres. A first order Banach bundle (C*-bundle) is also called *homogeneous*.

As a matter of terminology, if P is a property which bundles may or may not have, we say ξ has property P locally to mean there is some open cover $\mathcal{U}$ of the base space such that for each $U \in \mathcal{U}$, $\xi|U$ has property P. If ξ is any bundle over X, $A \subset X$, then $[A,\xi]$ denotes the set of homotopy classes of elements of $\Gamma(A,\xi)$, where a homotopy here must be a section at each stage. If Y is a space, we denote by $\mu_t : Y \to Y \times I$ the map $\mu_t(y) = (y,t)$, $t \in I = [0,1]$. If $s \in \Gamma(A,\xi)$, then $[s]$ denotes the equivalence class of s in $[A,\xi]$. It is easy to see that $[s_0] = [s_1]$ in $[A,\xi]$ if and only if there is $h \in \Gamma(A \times I, \xi \times I)$ with $\mu_0^*(h) = s_0$ and $\mu_1^*(h) = s_1$. It is also easy to see that if ξ is the product bundle with fibre F, then identifying each section with its principal part gives

$[A,\xi] = [A,F]$.

Of primary importance here will be the *bundle of maps*, $L(\xi,\zeta)$, where ξ is a locally trivial Banach bundle over X with finite dimensional fibre and ζ is any Banach bundle over X which is locally full [14]. The main fact here is that $L(\xi,\zeta)$ is itself a Banach bundle over X with a natural identification

$$\mathrm{Hom}_X(\xi,\zeta) = \Gamma(L(\xi,\zeta)),$$

that is also natural with respect to inducing. Thus, if $f : Y \to X$, then $f^*(L(\xi,\zeta)) = L(f^*(\xi),f^*(\zeta))$ and the diagram

$$\begin{array}{ccc} \mathrm{Hom}_X(\xi,\zeta) & \longrightarrow & \Gamma(L(\xi,\zeta)) \\ \downarrow f^* & & \downarrow f^* \\ \mathrm{Hom}_Y(f^*(\xi),f^*(\zeta)) & \longrightarrow & \Gamma(L(f^*(\xi),f^*(\zeta))) \end{array}$$

is commutative. We denote by $U(\xi,\zeta)$ the subbundle of $L(\xi,\zeta)$ consisting of the norm preserving linear maps. A member ϕ of $\Gamma(U(\xi,\zeta))$ is a Banach bundle isomorphism of ξ onto a Banach subbundle of ζ, a situation which we henceforth denote by $\phi : \xi < \zeta$. If ξ and ζ are also C*-bundles, then $C^*L(\xi,\zeta)$ denotes the subbundle of $L(\xi,\zeta)$ consisting of *-homomorphisms and $C^*U(\xi,\zeta)$ denotes the subbundle of $L(\xi,\zeta)$ consisting of injective *-homomorphisms. If ξ and ζ are in fact C_1^*-bundles, then $C_1^*U(\xi,\zeta)$ denotes the subbundle of $C^*U(\xi,\zeta)$ consisting of identity preserving (unital) *-homomorphisms. Likewise, if ξ and ζ are SC_0^*-bundles, then $C_0^*U(\xi,\zeta)$ denotes the subbundle of $C^*U(\xi,\zeta)$ consisting of the *trace preserving* *-homomorphisms. In general, $U(\xi,\zeta)$, $C^*L(\xi,\zeta)$, $C^*U(\xi,\zeta)$, $C_1^*U(\xi,\zeta)$, and $C_0^*U(\xi,\zeta)$ are neither Banach bundles nor fibre bundles. Of course, if ζ is also locally trivial with finite dimensional fibre, then all of these "associated bundles" will again be fibre bundles.

In certain situations, consistency of notation will make it convenient for us to define

$$SC_i^*U(\xi,\zeta) = C_i^*U(\xi,\zeta), \quad i = 0,1$$

when ξ and ζ are SC_i^*-bundles. We may therefore speak of $\mathcal{C}U(\xi,\zeta)$, where ξ and ζ are $\mathcal{C}$-bundles, ξ is homogeneous with finite dimensional fibre, and $\mathcal{C} \in \{C^*, C^*_1, SC_1^*, SC_0^*\}$. Set $\mathcal{C}U(\xi,\xi) = \mathcal{C}U(\xi)$. The sections of $\mathcal{C}U(\xi)$ are thus $\mathcal{C}$-bundle automorphisms of the homogeneous bundle ξ and $\mathcal{C}U(\xi)$ is a fibre bundle.

We will where convenient identify bundles over a single point with spaces. Since any Banach space can be considered a Banach bundle over a single point, if F and G are Banach spaces with F finite dimensional, then L(F,G) is still the Banach space of bounded linear maps. However, we can also speak of the space U(F,G), and if F and G are C^*-algebras, we have spaces $C^*U(F,G)$ and $C_1^*U(F,G)$, etc. As our main theorem will show, the space $C^*U(F,G)$ for F and G finite dimensional C^*-algebras is of the utmost interest for classifying second order C^*-bundles. For if ξ and ζ are homogeneous C^*-bundles with fibres F and G respectively, then $C^*U(\xi,\zeta)$ is locally trivial with fibre $C^*U(F,G)$.

Let $\mathcal{C} \in \{SC_0^*, SC_1^*, C_1^*, C^*\}$ and let X be a space. Suppose now that A is a closed subset of X and β_0 is a fixed $\mathcal{C}$-bundle over A, which is homogeneous with fibre F of finite dimension. Let ξ be a fixed $\mathcal{C}$-bundle over $X\backslash A$. Let $\mathcal{CB}_X(\beta_0,\xi)$ denote the set of isomorphism classes of $\mathcal{C}$-bundles ζ over X with $\zeta|A \cong \beta_0$ and $\zeta|X\backslash A \cong \xi$. Obviously $\mathcal{CB}_X(\beta_0,\xi)$ depends only on the isomorphism classes $[\beta_0]$ and $[\xi]$, so we can write $\mathcal{CB}_X(\beta_0,\xi) = \mathcal{CB}_X([\beta_0],[\xi])$. We can also form $\mathcal{CB}_{X\times I}(\beta_0 \times I,\ \xi \times I)$, and we define an equivalence relation $\underset{\mathcal{C}}{\approx}$ on $\mathcal{CB}_X(\beta_0,\xi)$ by saying $\zeta_0 \underset{\mathcal{C}}{\approx} \zeta_1$ or $[\zeta_0] \underset{\mathcal{C}}{\approx} [\zeta_1]$ for $[\zeta_0],[\zeta_1] \in \mathcal{CB}_X(\beta_0,\xi)$ if there is $[\zeta] \in \mathcal{CB}_{X\times I}(\beta_0 \times I,\ \xi \times I)$ with $\mu_i^*(\zeta) \cong \zeta_i$, $i = 0,1$. If $[\zeta_0] \underset{\mathcal{C}}{\approx} [\zeta_1]$, we shall say ζ_0 and ζ_1 are *$\mathcal{C}$-homotopic*.

Let $\Pi\mathcal{CB}_X(\beta_0,\xi) = \mathcal{CB}_X(\beta_0,\xi)/\underset{\mathcal{C}}{\approx}$. Thus our problem is to classify elements of $\Pi\mathcal{CB}_X(\beta_0,\xi)$ and then study the relation $\underset{\mathcal{C}}{\approx}$.

Let W be a fixed open neighborhood of A in X and β a homogeneous $\mathcal{C}$-bundle over W such that $\beta|A \cong \beta_0$. For simplicity of notation, we set $\mathcal{C}U(\beta,\xi) = \mathcal{C}U(\beta|W\backslash A,\ \xi|W\backslash A)$. Suppose $[\phi] \in [V\backslash A,\ \mathcal{C}U(\beta,\xi)]$ where V is an open neighborhood of A contained in W, or in the notation of [14], $V \in N(W,A)$. Using 6.2, we form the $\mathcal{C}$-bundle $b(\phi) = (\beta|V) \cup_\phi \xi$ over X, to get $[b(\phi)] \in \mathcal{CB}_X(\beta_0,\xi)$. Now we must determine the relation among such ϕ's which corresponds to the homotopy relation for $\mathcal{C}$-bundles in $\mathcal{CB}_X(\beta_0,\xi)$.

Let $\mathcal{CH}(\xi)$ denote the group of isotopy classes of ($\mathcal{C}$-bundle) automorphisms of ξ in analogy with [13, Definition 6.2]. If ϕ is an automorphism of ξ, then $[\phi]$ denotes the isotopy class of ϕ in $\mathcal{CH}(\xi)$. For $[\phi_0]$ and $[\phi_1]$ in $\mathcal{CH}(\xi)$ we have $[\phi_0] = [\phi_1]$ if and only if there is a $\mathcal{C}$-bundle automorphism ψ of $\xi \times I$ over $X \times I$ such that $\mu_0^*(\psi) = \phi_0$ and $\mu_1^*(\psi) = \phi_1$. Restriction defines a group homomorphism $\mathcal{CH}(\xi) \to \mathcal{CH}(\xi|V\backslash A)$ whose image we denote by $\mathcal{CH}(\xi) \mid V\backslash A$. Composition of $\mathcal{C}$-bundle maps defines a left action of $\mathcal{CH}(\xi|V\backslash A)$ on $[V\backslash A, \mathcal{C}U(\beta,\xi)]$ and thus also an action of the subgroup $\mathcal{CH}(\xi)|V\backslash A$. Likewise, $\mathcal{CH}(\beta|V\backslash A) = [V\backslash A, \mathcal{C}U(\beta)]$ acts on $[V\backslash A, \mathcal{C}U(\beta,\xi)]$ from the right giving an action of $\mathcal{CH}(\beta|V)|V\backslash A = [V, \mathcal{C}U(\beta)]|V\backslash A$ on $[V\backslash A, \mathcal{C}U(\beta,\xi)]$. Thus if G_* denotes the group obtained from the group G by reversing the order of multiplication, then

$$G(V) = (\mathcal{CH}(\xi)|V\backslash A) \times ([V,\mathcal{C}U(\beta)]|V\backslash A)_*$$

acts on $[V\backslash A, \mathcal{C}U(\beta,\xi)]$ from the left in the obvious manner, and in analogy with [14, page 307] we denote the resulting orbit set by $\mathcal{CO}_X(\beta,\xi,V)$, or simply $\mathcal{CO}_X(V)$ for simplicity. If $\phi \in \Gamma(V\backslash A,\mathcal{C}U(\beta,\xi))$, we denote by $\langle\phi\rangle$ its class in $\mathcal{CO}_X(V)$. If $V_1 \in N(V,A)$, and $i : V_1 \subseteq V$, then restriction to V_1 induces a natural map $i^* : \mathcal{CO}_X(V) \to \mathcal{CO}_X(V_1)$ and if $V_2 \in N(V_1,A)$ with $j : V_2 \subset V_1$, then $(i \circ j)^* = j^* \circ i^* : \mathcal{CO}_X(V) \to \mathcal{CO}_X(V_2)$. Thus the family $\{\mathcal{CO}_X(V) : V \in N(W,A)\}$ together with the maps induced by inclusion maps forms a direct system whose direct limit we denote

$$\mathcal{CO}_X(\beta,\xi) = \varinjlim_{V \in N(W,A)} \mathcal{CO}_X(\beta,\xi,V).$$

This is in exact analogy with the formulation for Hilbert bundles of [14, page 307]. We take this opportunity to point out that in the formulas given there, [14, page 307], there is a notational error: $\mathcal{O}(V)$ is $\mathcal{O}(\beta,\xi,V)$ or $\mathcal{O}(\beta|V,\xi,V)$ but *not* $\mathcal{O}(\beta|V,\xi|V,V)$, as should be clear from context, since $\xi|V$ makes no sense.

MAIN THEOREM. *If* $X \times I$ *is a paracompact Hausdorff space, then the correspondence* $\phi \to b(\phi)$ *lifts to a bijection*

$$(*) \qquad b : \mathcal{CO}_X(\beta,\xi) \to \Pi\mathcal{CB}_X(\beta_0,\xi).$$

COROLLARY. *Assume* X *is a metric space and* $\mathcal{C}$ *is either* SC_0^*, SC_1^*, *or* C^*. *Assume that* ξ *has locally finite order. In case* $\mathcal{C} = SC_0^*$, *assume that* ξ *is proper (see section seven). In case* $\mathcal{C} \in \{C^*, C_1^*\}$ *assume that* ξ *is also homogeneous with finite dimensional fibres. Then* $(*)$ *is actually a bijection*

$$(**) \qquad b : \mathcal{CO}_X(\beta,\xi) \to \mathcal{CB}(\beta_0,\xi).$$

In *theory* our main theorem classifies the *homotopy* classes of finite order C*-bundles having only finite dimensional fibres, since it can be used to classify the homotopy classes of these bundles of order n in terms of homogeneous bundles and those of order less than n. By proposition 7.1, the problem of finally classifying the isomorphism classes of these bundles is merely a problem of understanding their local structure sufficiently. For $\mathcal{C} = SC_0^*$ or $\mathcal{C} = SC_1^*$, we do in fact know the local structure sufficiently well as our corollary gives the classification of the *isomorphism* classes of bundles in this case.

The Corollary thus gives a homotopy theoretic classification of the members of $\mathcal{CB}_X(\beta_0,\xi)$. The proofs will be given in section eight. If we begin merely with β_0 and ξ, to obtain a classification we need to extend β_0 to a homogeneous $\mathcal{C}$-bundle over some neighborhood of A. If β_0 is of finite type, which means there is a finite open cover of A each of whose members over which β_0 is trivial, then as the group of the fibre bundle β_0 is a compact Lie group, from [28] we know there is a $\mathcal{C}$-bundle γ over a compact manifold M with $\beta_0 \cong f^*(\gamma)$ for some continuous map $f : A \to M$. As M is an absolute neighborhood retract, f extends to a map of some neighborhood of A into M and thus a $\mathcal{C}$-bundle β with $\beta|A \cong \beta_0$ exists to give the classification. If β_1 and β_2 are two such $\mathcal{C}$-bundles over $W_1 \in N(X,A)$, then as $\beta_1|A \cong \beta_2|A$ and $\mathcal{CU}(\beta_1,\beta_2)$ has the NEP by 3.5, it follows that there is $W \in N(W_1,A)$ with $\beta_1|W \cong \beta_2|W$. In particular, if A is either compact or finite dimensional, then the needed extension β of β_0 always exists uniquely up to isomorphism and we can write $CO_X(\beta_0,\xi) = CO_X(\beta,\xi)$ unambiguously. In the final sections we shall examine some interesting special cases where $CO_X(\beta_0,\xi)$ simplifies enough to permit calculation of $CO_X(\beta_0,\xi)$.

Because it is so often convenient to identify bundle maps as sections of other bundles, we find it useful to denote by $\phi(x) : \xi(x) \to \zeta(f(x))$, $x \in X$, the map defined on the fibre $\xi(x)$, whenever $\phi : \xi \to \zeta$ is a bundle map covering $f : X \to Y$. We denote by $\xi^{(d)}$ the bundle obtained from the bundle ξ on replacing total space and base space topologies with discrete topologies, and we say $\phi : \xi \to \zeta$ is a *pre-bundle map* covering $f : X \to Y$ if $\phi : \xi^{(d)} \to \zeta$ is a bundle map. We say $\phi : \xi \to \zeta$ is a *pre-Banach* (C*-) *bundle map* covering $f : X \to Y$ if ϕ is a pre-bundle map of Banach bundles (C*-bundles) such that for each $x \in X$, $\phi(x) : \xi(x) \to \zeta(f(x))$ is a bounded linear map (*-homomorphism). In case $\phi = (h;f)$ is a pre-bundle map or bundle map, we will often identify ϕ with the map h of total spaces, where convenient. We define $\Pi(\xi) = \Gamma(\xi^{(d)})$, so that

$$\Pi(\xi) = \prod_{x\in X} \xi(x),$$

if ξ is a bundle over X. We shall call the members of $\Pi(\xi)$ *selections* of ξ. A *section* of ξ is thus a *continuous selection*. For $A \subset X$ And ξ a bundle over X, set $\Pi(A,\xi) = \Pi(\xi|A)$.

If ξ is any bundle over X, if $S \subset \Pi(\xi)$ and $x \in X$, then $S(x)$ denotes the set $S(x) = \{s(x) : s \in S\}$, and if $A \subset X$, then $S|A = \{s|A : s \in S\} \subset \Pi(A,\xi)$. We recall that $S \subset \Gamma(\xi)$ is said to be *full for* ξ provided for each $x \in X$, $S(x) = \xi(x)$. If ξ is a Banach bundle over X, then $S \subset \Gamma(\xi)$ is said to be *total* for ξ if for each $x \in X$, $S(x)$ spans a dense linear subspace of $\xi(x)$. We say the bundle ξ is *full* if $\Gamma(\xi)$ is full for ξ. From [18, Prop.1.5] we know that if there is $S \subset \Gamma(\xi)$ total for the Banach bundle ξ, then ξ is full (notice that even the Hausdorff assumption on the spaces involved there is unnecessary).

Because of the usefullness of sections in the theory, we make the following assumption

STANDING HYPOTHESIS. *All Banach bundles (and* C*-*bundles) are assumed to be full.*

Actually, from [11,31] every Banach bundle over a paracompact Hausdorff space is full. In our main applications, the base spaces are all in fact metrizable, so in that case, the standing hypothesis is actually redundant. Thus, if we claim a bundle ξ produced in some fashion or construction is a Banach bundle or C*-bundle, then we are in fact also claiming it is full. However, this will usually be clear from the construction.

Before concluding this section, we record a few basic results for future reference.

PROPOSITION 1.1. *Let* $\phi : \xi \to \zeta$ *be a Banach* (C*-) *bundle map covering a map* $f : X \to Y$, *and let* $\psi : f^*(\zeta) \to \zeta$ *be the induced Banach* (C*-) *bundle map. Then there is* $\theta : \xi \to f^*(\zeta)$ *over* X *a unique Banach* (C*-) *bundle map such that* $\phi = \psi \circ \theta$. *Moreover,* ϕ *is a strong Banach* (C*-) *bundle map if and only if* $\theta : \xi \cong f^*(\zeta)$.

For the proof of 1.1, we refer the reader to [13, 1.1,1.3,1.4, 1.5] and [9, Prop. 6]. Notice that $\phi : \xi \to \zeta$ is a strong C*-bundle map covering f (i.e. C*-bundle map which is strong as a Banach bundle map) if and only if for each x in the base space of ξ, $\phi(x) : \xi(x) \to \zeta(f(x))$ is a *-isomorphism of C*-algebras, as then $\phi(x)$ is automatically isometric.

If ξ is a Banach bundle over X and s is a *selection* of ξ, then $|s|$ denotes the real-valued function $x \to \|s(x)\|$ on X. Likewise, if $\phi : \xi \to \zeta$ is a pre-Banach bundle map covering $f : X \to Y$, then $|\phi|$ is the real-valued function on X given by $|\phi|(x) = \|\phi(x)\|$, $x \in X$.

PROPOSITION 1.2. *Let* $\phi : \xi \to \zeta$ *be a pre-Banach bundle map covering the (continuous) map* $f : X \to Y$. *In order that* ϕ *be a Banach bundle map, it is necessary and sufficient that*

(1) *the function* $|\phi|$ *is locally bounded on* X, *and*

(2) *there is* S *total for* ξ *such that for each* $s \in S$, *the function* $\phi \circ s$ *is a continuous map from* X *to the total space of* ζ.

Proof. $\phi : \xi^{(d)} \to \zeta$ is a bundle map so by the *factorization property*, [13, Prop. 1.1], there is $\theta : \xi^{(d)} \to f^*(\zeta)$ over X unique such that $\phi = \psi \circ \theta$, where ψ denotes the Banach bundle map induced by f, $\psi : f^*(\zeta) \to \zeta$. Thus, it suffices to show θ is a Banach bundle map. For each $x \in X$, $\psi(x)$ is an isometric isomorphism, so $|\phi| = |\theta|$, hence (1) holds if and only if $|\theta|$ is locally bounded on X. Likewise condition (2) is equivalent to the existence of S total for ξ such that for each $s \in S$, $\theta_*(s) \in \Gamma(f^*(\zeta))$, hence by [9, Prop. 5], θ is a Banach bundle map if and only if (1) and (2) hold, but since $\phi = \psi \circ \theta$, by 1.1 we see ϕ is a Banach bundle map if and only if θ is.

PROPOSITION 1.3. *Let* $f : X \to Y$ *and* $g : Y \to Z$ *be continuous maps, and* $\psi : \xi \to \zeta$ *a strong Banach bundle map covering* g. *If* $\phi : \beta \to \zeta$ *is a Banach bundle map covering* $g \circ f$, *then there is* $\theta : \beta \to \xi$ *a unique Banach bundle map covering* f *such that* $\phi = \psi \circ \theta$.

Proof. Let $\psi^* : f^*(\xi) \to \xi$ be the bundle map induced by f. Then $\psi \circ \psi^*$ is a strong Banach bundle map covering $g \circ f$ so there is $\theta_1 : \beta \to f^*(\xi)$ over X unique such that $\psi \circ \psi^* \circ \theta_1 = \phi$. Putting $\theta = \psi^* \circ \theta_1$, we have θ covers f and $\psi \circ \theta = \phi$. The uniqueness is just as easy to prove.

Let $\phi : \xi \to \zeta$ be a strong Banach (C*-) bundle map covering a map $f : X \to Y$. Then $f^* : \Gamma(\zeta) \to \Gamma(f^*(\zeta))$ is linear and carries $\Gamma_b(\zeta)$ into $\Gamma_b(f^*(\zeta))$, so by 1.1 we obtain $\phi^* : \Gamma_b(\zeta) \to \Gamma_b(\xi)$ a bounded linear map with $\|\phi^*\| \le 1$, essentially by factoring sections, so $\phi^*(s) \in \Gamma_b(\xi)$ is unique such that $\phi \circ (\phi^*(s)) = s \circ f$, if $s \in \Gamma_b(\zeta)$. Moreover, if ξ and ζ are C*-bundles and ϕ is a strong C*-bundle map, then ϕ^* is a *-homomorphism. We can define a Banach subbundle η of ζ, if Y is paracompact Hausdorff, by setting

$$\eta(y) = \begin{cases} 0, & y \in \overline{f(X)} \\ \zeta(y), & y \in Y \setminus \overline{f(X)}, \end{cases}$$

as in [14, page 297]. We state the next result for future reference.

PROPOSITION 1.4. *With* η *as above,* $\Gamma_b(\eta) = \operatorname{Ker} \phi^*$.

Proof. The result is obvious.

Added in proof: If $X : E \to G$ is any covariant functor, we can think of the objects of E as generalized bundles and the morphisms of E as bundle maps, and if $\xi \in E$ we think of $X(\xi)$ as the base object ξ. Likewise, if $\phi : \xi \to \zeta$ is a morphism of E, we say ϕ covers $X(\phi)$. Obviously much of the elementary terminology of bundle theory can be extended to this level of generality. In particular, $\psi : \xi \to \zeta$ is called X-cartesian if the analogue of 1.1 holds for ψ, and is called X-hypercartesian if the stronger 1.3 holds for ψ. The composition of hypercartesian maps is again hypercartesian, but the composition of cartesian maps need not be. X is called a *fibration* if the composition of cartesian maps is cartesian and if for any $\zeta \in E$ and any $f : S \to X(\zeta)$, there is some cartesian ψ covering f with codomain ζ. The proof of 1.3 then shows every cartesian map is hypercartesian, and 1.1 shows that our various Banach bundle categories are fibrations over the category of topological spaces. In [12] this author was led to define the notion of fibred category and called it a general bundle theory, unaware of the literature concerning fibred categories already in existence [20]. We will say more about generalizations of our theory in section 8.

2. THE NEIGHBORHOOD EXTENSION PROPERTY

We say a bundle ξ over X has the *neighborhood extension property* in $Y \subseteq X$, denoted NEP(Y), if it satisfies

NEP(Y): for every set $A \subset Y$, with A closed in X, and every $s \in \Gamma(A,\xi)$, there is U open in Y with $t \in \Gamma(U,\xi)$ which extends s.

We say ξ has the NEP if it has the NEP(X). Notice that if ξ has the NEP over $Y \subseteq X$, then ξ has the NEP(Y), but the converse may not hold. Moreover, it is clear that if $Y_2 \subseteq Y_1 \subseteq X$ and if ξ has the NEP(Y_1), then ξ has the NEP(Y_2). However, if ξ has the NEP over Y_1, ξ may fail to have the NEP over Y_2. The NEP is important as a first step toward obtaining global extensions as shown by the theory developed in [10]. There, the NEP is used as the hypothesis essentially in proving global extension properties of bundles follow from local global extension properties. We shall need the analogous result for the NEP itself as expressed in the theorem below. We say ξ has the NEPL to mean there is an open cover $\mathcal{U}$ of X such that for each $U \in \mathcal{U}$, ξ has the NEP(U).

THEOREM 2.1. *Let* X *be a paracompact Hausdorff space and* ξ *a bundle over* X *which has the* NEPL. *Then* ξ *has the* NEP *globally. In particular, if* ξ *locally has the* NEP, *then* ξ *has the* NEP.

<u>Proof</u>. If $\mathcal{V}$ is a family of disjoint open subsets of X such that for each $U \in \mathcal{V}$, ξ has the NEP($\mathcal{U}$), then ξ has the NEP($\cup\mathcal{V}$). Let $\mathcal{U}$ be an open cover of X such that for each $U \in \mathcal{U}$, ξ has the NEP(U). As $\mathcal{U}$ has a σ-discrete open refinement (because X is paracompact Hausdorff) we may assume on passing to a refinement if necessary that $\mathcal{U} = \bigcup_{i=1}^{\infty} \mathcal{U}_i$ where each $\mathcal{U}_i$ is a disjoint collection of open sets. Let $U_i = \cup\mathcal{U}_i$. Then $\{U_1,U_2,U_3,\ldots\}$ is an open cover of X and ξ has the NEP(U_i), $i = 1,2,3,\ldots$. Now suppose A is closed in X, $s \in \Gamma(A,\xi)$. There is an open cover $\{V_1,V_2,V_3,\ldots\}$ of X such that for each i, $\overline{V}_i$, the closure of V_i, is contained in U_i, because X is paracompact. Set $V_0 = S_0 = \emptyset$. $s_0 = s$, and $T_0 = A$. Proceeding inductively, assume that

open sets $S_1, S_2, \ldots, S_n$, closed sets $T_0 \subset T_1 \subset T_2 \subset \ldots \subset T_n$, sections $s_i \in \Gamma(T_i, \xi)$, $1 \le i \le n$, have been found such that for each i, $1 \le i \le n$, we have $\overline{S}_i \subset U_i$, $T_{i-1} \cap \overline{V}_i \subset S_i$, $T_i = T_{i-1} \cup (\overline{S}_i \cap \overline{V}_i)$, and $s_i | T_{i-1} = s_{i-1}$, $1 \le i \le n$. The proof for step $n = 1$ to begin this process we can safely leave to the reader as it is essentially the same as the argument for the inductive step which we now give. As $T_n \cap \overline{V}_{n+1}$ is closed in X, ξ has the $\mathrm{NEP}(U_{n+1})$, and X is a normal space, there is S_{n+1} open in X with $\overline{S}_{n+1} \subset U_{n+1}$, $T_n \cap \overline{V}_{n+1} \subset S_{n+1}$ and $s'_{n+1} \in \Gamma(\overline{S}_{n+1}, \xi)$ such that

$$s'_{n+1} \mid T_n \cap \overline{V}_{n+1} = s_n \mid T_n \cap \overline{V}_{n+1} .$$

As $T_n \cap (\overline{S}_{n+1} \cap \overline{V}_{n+1}) = T_n \cap \overline{V}_{n+1}$, there is $s_{n+1} \in \Gamma(T_{n+1}, \xi)$ unique such that $s_{n+1} | T_n = s_n$ and $s_{n+1} | \overline{S}_{n+1} \cap \overline{V}_{n+1} = s'_{n+1} | \overline{S}_{n+1} \cap \overline{V}_{n+1}$, where $T_{n+1} = T_n \cup (\overline{S}_{n+1} \cap \overline{V}_{n+1})$. We thus obtain a sequence (S_i) of open sets, an increasing sequence (T_i) of closed sets, and a sequence of sections (s_i) with $s_i \in \Gamma(T_i, \xi)$, $1 = 1,2,3,\ldots$, such that for each i, $T_i = T_{i-1} \cup (\overline{S}_i \cap \overline{V}_i)$, $T_{i-1} \cap \overline{V}_i \subset S_i$, $s_i | T_{i-1} = s_{i-1}$. Set $W = \bigcup_{n=1}^{\infty} (S_i \cap V_i)$. Then W is open, and $W \subset \bigcup_{n=1}^{\infty} T_n$, hence there is a unique function $\tilde{s}$ on W such that for each $n \ge 1$, $\tilde{s} | S_n \cap V_n = s_n | S_n \cap V_n$, and as each $S_n \cap V_n$ is open, it follows that $\tilde{s} \in \Gamma(W, \xi)$. Also, as $A = T_0 \subset T_1 \subset T_2, \ldots$, it follows that for each n,

$$A \cap V_n \subset T_{n-1} \cap V_n \subset T_{n-1} \cap \overline{V}_n \subset S_n$$

and hence $A \cap V_n \subset S_n \cap V_n$. But, $\{V_1, V_2, V_3, \ldots\}$ covers X, so $A \subset W$. As $\tilde{s} \in \Gamma(W, \xi)$, $A \subset W$, and for each i, s_i extends s, the definition of $\tilde{s}$ guarantees that $\tilde{s} | A = s$. The proof is now complete.

Consider now a better extension property. We say the bundle ξ over X satisfies the *absolute extension property* (AEP) if it satisfies the following condition:

(AEP) for every closed set $A \subset X$ and every $s \in \Gamma(A, \xi)$, there is $t \in \Gamma(X, \xi)$ with $t | A = s$.

The (AEP) is clearly a stronger property than the (SEP) formulated in [10]. Combining the section extension theorem, [10, Theorem 2.7], with 1.1 gives the next interesting fact.

COROLLARY 2.2. *Let* ξ *be a bundle over the paracompact Hausdorff space* X *which locally has the* AEP. *Then* ξ *has the* AEP.

Proof. Since the AEP implies the NEP, by 2.1, ξ has the NEP globally. As the AEP implies the SEP as formulated in [10, definitions 2.1 and 2.2], ξ has the SEP locally and hence globally by [10, theorem 2.7]. Since the space X is normal Hausdorff, the SEP and the NEP together obviously imply the AEP, so ξ has the AEP globally, completing the proof.

We shall not need 2.2 until section 9, but 2.1 will be of the utmost importance, for we shall see that because of 2.1, the bundle $C^*U(\xi,\zeta)$ has the NEP when ξ is a homogeneous C*-bundle with finite dimensional fibre and ζ is any C*-bundle. This fact is crucial for our classification theory.

Added in proof:

It is interesting that the conclusion of 2.1 is known for the case where ξ is the space etalé of a sheaf. For instance see 3.3.1 of Chapter II of [32] or [33, 2.2.1]. However, in that case the proof is considerably simplified by the fact that two sections of a sheaf agree on some neighborhood of any point at which they agree. If Γ_ξ denotes the sheaf of germs of sections of ξ, then $\Gamma_\xi(S) = \Gamma(S,\xi)$ for S open in X, but not for S closed in X until 2.1 is applied.

3. THE EXTENSION OF AN EMBEDDING

Our aim in this section is to show that if β and ξ are C*-bundles over the same paracompact Hausdorff base space and if β is homogeneous with finite dimensional fibre, then $C^*U(\beta,\xi)$ has the NEP as defined in the previous section. To accomplish this, we will have to apply some C*-algebra theory, for the notation of which we refer to [8]. First we prove some lemmas.

LEMMA 3.1. *Let* ξ *be a* C*-*bundle over a space* X, *and let* $a \in \Gamma_b(\xi)$ *be a normal* $(a^*a = aa^*)$ *section. Then* $Sp'(a)$ *is the closure of* $\bigcup_{x \in X} Sp'(a(x))$.

Proof. Algebraic considerations alone make it clear that if λ belongs to $Sp'(a(x))$ for some $x \in X$, then λ belongs to $Sp'(a)$. Thus, if K denotes the union of the collection $\{Sp'(a(x)) : x \in X\}$, then $K \subset Sp'(a)$. As $Sp'(a)$ is compact, it suffices to show that $Sp'(a)$ is contained in the closure $\overline{K}$, of the set K. Suppose $\lambda \in Sp'(a)$ does not belong to $\overline{K}$. Choose a continuous function $f : \mathbb{C} \to [0,1]$ such that $f(\lambda) = 1$ and with f vanishing on $\overline{K}$. Since $0 \in K$, it follows that $f(0) = 0$, so that $f(a)$ can be formed in the C*-algebra $\Gamma_b(\xi)$. Point evaluations are *-homomorphisms, so for each $x \in X$ we must have $f(a)(x) = f(a(x)) = 0$, hence $f(a) = 0$. On the other hand, by the spectral mapping theorem, $1 = f(\lambda)$ must belong to $Sp'(f(a))$ so that $f(a) \neq 0$, a contradiction which completes the proof of the lemma.

For the next three lemmas we shall assume that ξ *is a* C*-*bundle over* X, *that* A *is a subset of* X, *and that each* $s \in \Gamma(A,\xi)$ *extends to a section of* $\xi|X_s$ *for* X_s *some open neighborhood of* A *in* X *depending on the particular* $s \in \Gamma(A,\xi)$ *to be extended.* We point out that if A consists of a single point or if A is closed and X is paracompact Hausdorff [13, proposition 1.7], then our hypothesis is satisfied by taking $X_s = X$ for every section $s \in \Gamma(A,\xi)$.

LEMMA 3.2. *Let* $s_1,\ldots,s_n$ *be pairwise orthogonal projection fields* $(s_i = s_i^*,\ s_i^2 = s_i,\ 1 \le i \le n)$ *of* $\xi|A$. *Then there is* U *open in* X *with* $A \subset U$, *and pairwise orthogonal projection fields* $t_1,\ldots,t_n$ *of* $\xi|U$ *such that* $t_i|A = s_i,\ 1 \le i \le n$.

For the proof of 3.2, we have adapted to our situation a simple technique due to Thayer [30], which does away with the need to generalize the complicated argument in [19, lemma 3.1] and [17, lemma 2.5].

Proof. Set $K = \{0\} \cup \{k^{-1} : 1 \le k \le n\}$ and define $s \in \Gamma_b(A,\xi)$ by the formula

$$s = \sum_{k=1}^{n} \frac{1}{\sqrt{k}} s_k .$$

Then $s = s^* \ge 0$ and $Sp'(s) = K^{\frac{1}{2}} = \{0\} \cup \{k^{\frac{1}{2}} : 1 \le k \le n\}$. We can choose $a \in \Gamma(X_s,\xi)$ with $a|A = s$. Since the norm is continuous on the total space and $s \in \Gamma_b(A,\xi)$, a must be bounded on some open neighborhood of A which we may assume is X_s after making X_s smaller if necessary. Thus $a \in \Gamma_b(X_s,\xi)$. Set $b = a^*a$. Then $b = b^* \ge 0$ and $b|A = s^2$. Thus b is a normal section of $\xi|X_s$ and $Sp'(b(x)) = K$, $x \in A$. Let W be a closed neighborhood of K in $\mathbb{C}$ such that K is a retract of W and let $r : W \to K$ be the retraction. By [8, 10.3.6], the set of points $x \in X_s$ for which $Sp'(b(x))$ is contained in the interior of W must be an open set U in X_s. Thus $A \subset U$ and U is open in X. By 3.1, $Sp'(b|U)$ is contained in the closed set W hence we can form $r(b|U)$ in the C*-algebra $\Gamma_b(U,\xi)$. Set $t = r(b|U)$. Then $Sp'(t) = K$ and as r is a retraction, $t|A = b|A = s^2 = \sum_{k=1}^{n} \frac{1}{k} s_k$. For each k, $1 \le k \le n$, let f_k denote the characteristic function of the single point set $\{k^{-1}\}$ in the set K. Then f_k is continuous on K so we can define $t_k = f_k(t)$. Then $t_1,t_2,\ldots,t_n$ are pairwise orthogonal projection fields and for each positive integer $k \le n$ we have $t_k|A = s_k$. The proof is now complete.

LEMMA 3.3. *Let* q_1 *and* q_2 *be projection fields of* ξ. *Let* $u \in \Gamma(A,\xi)$ *such that* $u^*u = q_1|A$ *and* $uu^* = q_2|A$. *Then there is* U *open in* X *with* $A \subset U$ *and* $v \in \Gamma(U,\xi)$ *such that* $v^*v = q_1|U$, $vv^* = q_2|U$, *and* $v|A = u$.

Proof. We shall use the technique of [19, lemma 3.2] where the proof is given in case A is a single point and X is locally compact Hausdorff.

We can choose $t' \in \Gamma(X_u,\xi)$ such that $t'|A = u$. Set $t = q_2 t' q_1$. Then $q_2 t = t = tq_1$, and $t|A = u$. Since norm is continuous on the total space of ξ and u is bounded, after replacing X_u by a smaller open neighborhood of A and restricting our sections to this new X_u, we may assume $t \in \Gamma_b(X_u,\xi)$. We can then form $h = (t^*t)^{\frac{1}{2}}$ in the C*-algebra $\Gamma_b(X_u,\xi)$. As restriction $\Gamma_b(X_u,\xi) \to \Gamma_b(A,\xi)$ is a *-homomorphism, we have

$$h|A = (u^*u)^{\frac{1}{2}} = q_1|A.$$

Since $t = tq_1$, it follows that $t^*t \in q_1\Gamma_b(X_u,\xi)q_1$ and therefore $h \in q_1\Gamma_b(X_u,\xi)q_1$, so $h = q_1h = hq_1$.

Since $h-q_1$ vanishes on A, after again making X_u smaller if necessary, we may assume $\|h-q_1\| < \frac{1}{4}$. Since h and q_1 are commuting self-adjoint elements of $\Gamma_b(X_u,\xi)$ and since q_1 is a projection, it follows that $Sp'(h)$ is contained in $[-\frac{1}{4}, \frac{1}{4}] \cup [\frac{3}{4}, \frac{4}{3}]$. Let $f : \mathbb{R} \to [0,\frac{4}{3}]$ be any continuous function such that $f|[-\frac{1}{4}, \frac{1}{4}] = 0$ and $f(r) = 1/r$, $|r-1| \leq 1/3$. Let $i : \mathbb{R} \to \mathbb{R}$ be the identity map. Then $i \cdot f$, the product of i and f in $C(\mathbb{R})$ is constant with value 0 on $[-\frac{1}{4}, \frac{1}{4}]$ and constant with value 1 on $[\frac{3}{4}, \frac{4}{3}]$. Thus $hf(h)$ is a projection. Let $w = tf(h)$. Since $h|A = q_1|A$ and restriction mapping $\Gamma_b(X_u,\xi) \to \Gamma_b(A,\xi)$ is a *-homomorphism, it follows that $f(h)|A = q_1|A$ and thus $hf(h)|A = q_1|A$. Computing now, $w^*w = f(h)t^*tf(h) = (f(h)h)^2 = f(h)h$, since $f(h)h$ is a projection, thus w^*w is a projection with $w^*w|A = q_1|A$. On the other hand, since h belongs to $q_1\Gamma_b(X_u,\xi)q_1$, it follows that $w^*w = hf(h)$ does also and therefore $q_1 - w^*w$ is a projection field vanishing on A. Making X_u smaller if necessary we can then assume $\|q_1 - w^*w\| \leq \frac{1}{2}$, and as $q_1 - w^*w$ is a projection we now have $q_1 = w^*w$. Since w^*w is a projection, ww^* is also a projection in $\Gamma_b(X_u,\xi)$, but $ww^* = tf(h)f(h)t^*$. As $q_2t = t$, it follows that ww^* is a projection in $q_2\Gamma_b(X_u,\xi)q_2$ and thus $q_2 - ww^*$ is a projection field. Again using the fact that restriction $\Gamma_b(X_u,\xi) \to \Gamma_b(A,\xi)$ is a *-homomorphism, we have $w|A = uf(q_1|A) = u(q_1|A) = u$. Thus $q_2 - ww^*$ is a projection field vanishing on A so there is finally an open neighborhood U of A in X_u with $\|(q_2 - ww^*)|U\| \le \frac{1}{2}$, and therefore with $q_2|U = (ww^*)|U$. Setting $v = w|U$ completes the proof.

LEMMA 3.4. *Let* F *be a finite dimensional* C*-*algebra and* $\beta = \varepsilon(X,F)$, *the product* C*-*bundle over* X *with fibre* F. *Let* $\phi : \beta|A < \xi|A$ *be a* C*-*bundle embedding. Then there is* U *open in* X *with* $A \subset U$ *and* $\psi : \beta|U < \xi|U$ *a* C*-*bundle embedding with* $\psi|A = \phi$.

Proof. Choose a system $(e_{ij}^{(k)})$ of matrix units for F so that

$$(*) \qquad e_{ij}^{(k)*} = e_{ji}^{(k)}, \; e_{ij}^{(k)} e_{mn}^{(\ell)} = \begin{cases} 0 & , k \neq \ell \text{ or } j \neq m \\ e_{in}^{(k)} & , k = \ell \text{ and } j = m \end{cases}$$

where $1 \le k \le r$ and for each k, $e_{ij}^{(k)}$ is defined for $1 \le i,j \le r_k$. Thus $F = \prod_{k=1}^{r} M(r_k)$. Using ϕ and (*) we obtain a family $(\tilde{e}_{ij}^{(k)})$ of matrix units in $\Gamma_b(A,\xi)$. Using a finite number of applications of 3.2 and 3.3, we find an open neighborhood U of A in X and a system $(\bar{e}_{ij}^{(k)})$ of matrix units in $\Gamma_b(U,\xi)$ satisfying (*) such that for each i,j,k, $\bar{e}_{ij}^{(k)}|A = \tilde{e}_{ij}^{(k)}$. Let $(w_{ij}^{(k)})$ be the basis of $L(F,\mathbb{C})$ dual to $(e_{ij}^{(k)})$. The family $(\tilde{e}_{ij}^{(k)})$ is chosen so that

$$\phi(x,b) = \sum_{i,j,k} w_{ij}^{(k)}(b)\tilde{e}_{ij}^{(k)}(x), \quad (x,b) \in A \times F.$$

We can thus define $\psi : \beta|U \to \xi|U$ by

$$\psi(x,b) = \sum_{i,j,k} w_{ij}^{(k)}(b)\bar{e}_{ij}^{(k)}(x).$$

Then ψ is obviously a C*-bundle map which is injective on fibres and $\psi|A = \phi$. Thus $\psi : \beta|U < \xi|U$ is a C*-bundle embedding extending ϕ and the proof is complete.

THEOREM 3.5. *Let* β *and* ξ *be* C*-*bundles over the paracompact Hausdorff space* X, *and suppose* β *is homogeneous with fibre* F *a finite dimensional* C*-*algebra. Then* $C^*U(\beta,\xi)$ *has the* NEP *(defined in the preceeding section). Moreover,* β *is a locally trivial* C_1^*-*bundle and if* ξ *is also a* C_1^*-*bundle, then* $C_1^*U(\beta,\xi)$ *has the* NEP.

Proof. Let ε be the product bundle with fibre F over X. Then ε is a C_1^*-bundle (every finite dimensional C*-algebra has identity) and by lemma 3.4, $C^*U(\varepsilon,\beta)$ has the NEP. Since any *-isomorphism of C*-algebras having identity must be identity preserving and since the local sections of $C^*U(\varepsilon,\beta)$ must in fact be local isomorphisms, it follows that β is a locally trivial C_1^*-bundle. But then $C^*U(\varepsilon,\xi)$ and $C^*U(\beta,\xi)$ are locally isomorphic. Thus, by lemma 3.4, $C^*U(\beta,\xi)$ has the NEP locally so that by theorem 2.1, $C^*U(\beta,\xi)$ has the NEP. Suppose now that ξ is also a C_1^*-bundle. Let A be a closed subset of X and $u \in \Gamma(A, C_1^*U(\beta,\xi))$. Since $C^*U(\beta,\xi)$ has the NEP, there is U open in X with $v \in \Gamma(U, C^*U(\beta,\xi))$ such that $v|A = u$. Let 1 denote the identity sections of either β or ξ, and view v as a C*-bundle map $v : \beta|U \to \xi|U$. Then $v \circ 1 = q$ is a projection field of ξ, hence $1-q$ is also a projection field of ξ. As $v|A = u$ is identity preserving, $1-q$ vanishes on A. But, since $1-q$ is a projection field, $f = |1-q| : U \to \mathbb{R}$ is continuous and takes on only the values 0 or 1. Hence $V = f^{-1}(0)$ is open and $v|V \in \Gamma(V, C_1^*U(\beta,\xi))$ with $[v|V]|A = u$. The proof is complete.

PROPOSITION 3.6. *Any homogeneous* C*-*bundle having a finite dimensional fibre* F *is a fibre bundle with group* $\mathrm{Aut}(F)$, *the group of *-automorphisms of* F. *In particular, if the fibre is* $M(n)$, *then the group of the bundle is* $PU(n) = U(n)/S^1$.

Proof. This is an immediate consequence of lemma 3.4.

We remark that as should be expected 3.6 implies Fell's theorem [19] that a homogeneous C*-algebra with finite rank is the algebra of cross-sections vanishing at infinity of a fibre bundle with group $PU(n)$ and fibre $M(n)$ over a locally compact Hausdorff space.

COROLLARY 3.7. *If* ξ *is a homogeneous* C*-*bundle with finite dimensional fibre, then* ξ *is a* C_1^*-*bundle. In particular, if the fibre is* $\mathbb{C}$, *then* ξ *is trivial with unique trivialization given by the identity section.*

Proof. Since any finite dimensional C*-algebra F must have identity, if F is the fibre of ξ, then ξ is a fibre bundle with group $\mathrm{Aut}(F)$. As every *-automorphism of F must be unital, it follows that the identity section of ξ is continuous and the proof is complete.

LEMMA 3.8. *Let* ξ *be an* SC_0^*-*bundle over* X, *and suppose* e *is a projection field of* ξ. *Then the set of points where* e *has rank one is open in* X.

Proof. Let R denote the set in question. Suppose $x_0 \in R$. There is V open in X with $x_0 \in V$ and $q \in \Gamma(V,\xi)$ a rank one projection field. There is $u \in \xi(x_0)$ with $u^*u = q(x_0)$ and $uu^* = e(x_0)$. By lemma 3.3, there is an open neighborhood W of x_0 contained in V and a section $v \in \Gamma(W,\xi)$ such that $v^*v = q|W$ and $vv^* = e|W$. It follows that $W \subset R$, hence R must be open.

PROPOSITION 3.9. *Let* X *be a paracompact Hausdorff space and* β *a homogeneous* C^**-bundle over* X *with fibre* $M(n)$. *Let* ξ *be an* SC_0^**-bundle over* X. *Then* β *is an* SC_0^**-bundle and* $C_0^*U(\beta,\xi)$ *has the* NEP.

Proof. Since β is a locally trivial SC^*-bundle it is obviously an SC_0^*-bundle. Let A be a closed subset of X and ϕ a section in $\Gamma(A,C_0^*U(\beta,\xi))$. By theorem 3.5, there is U open in X and containing A together with a section ψ of $\Gamma(U,C^*U(\beta,\xi))$ such that $\psi|A = \phi$. Let T be the set of all x in U such that $\psi(x)$ is trace preserving. We now observe that if $\pi : M(n) \to C$ is an injective *-homomorphism of simple dual C^*-algebras and if there is a rank one projection $e \in M(n)$ such that $\pi(e)$ is rank one in C, then π is trace preserving. Let $x_0 \in T$. There is V an open neighborhood of x_0 contained in U and $e \in \Gamma(V,\beta)$ a rank one projection field of $\beta|V$. Then $\psi \circ e \in \Gamma(V,\xi)$ is a projection field and has rank one at x_0 because $x_0 \in T$. It follows that there is an open neighborhood W of x_0 contained in V such that $\psi \circ e|W$ is a rank one projection field. Thus $\psi(x)$ is trace preserving for each $x \in W$ so $W \subset T$. We conclude that T is open, but $\psi|A = \phi \in \Gamma(A,C_0^*U(\beta,\xi))$ so that $A \subset T$, hence $\psi|T \in \Gamma(T,C_0^*U(\beta,\xi))$ and $\psi|T$ extends ϕ. This completes the proof that $C_0^*U(\beta,\xi)$ has the NEP.

Suppose now that ξ and ζ are homogeneous C^*-bundles over X with finite dimensional fibres F and G respectively, and that $\phi : \xi < \zeta$. We say $\phi : \xi < \zeta$ is *trivial* if there is an injective *-homomorphism $\phi_0 : F < G$, and trivializations $\theta : \xi \cong \varepsilon(X,F)$ and $\psi : \zeta \cong \varepsilon(X,G)$ such that

$$\begin{array}{ccc} \xi & \xrightarrow{\theta} & \varepsilon(X,F) \\ \phi\big\downarrow & & \big\downarrow\varepsilon(X,\phi_0) \\ \zeta & \xrightarrow{\psi} & \varepsilon(X,G) \end{array}$$

is a commutative diagram. We then call the pair (θ,ψ) a *trivialization* of ϕ. Our need to prove that such a ϕ is always locally trivial requires the following development. Since $C^*U(\xi,\zeta)$ is locally trivial with fibre $C^*U(F,G)$, our main theorem shows the space $C^*U(F,G)$ to be of primary interest for classifying second order C^*-bundles. First notice that composition of maps defines a left action of $\mathrm{Aut}(G)$ on $\mathrm{Hom}(F,G)$ under which $C^*U(F,G)$ is invariant.

LEMMA 3.10. *Let* F *and* G *be finite dimensional* C^**-algebras. Then* $C^*U(F,G)$ *is a compact open subset of* $\mathrm{Hom}(F,G)$ *and both these spaces are compact manifolds. Moreover,* $\mathrm{Hom}(F,G)/\mathrm{Aut}(G)$ *and* $C^*U(F,G)/\mathrm{Aut}(G)$*, the resulting orbit spaces, are finite discrete spaces.*

Proof. Since $C^*U(F,G) \subset \mathrm{Hom}(F,G)$ and $\mathrm{Hom}(F,G)$ is obviously compact, it suffices to show $C^*U(F,G)$ is closed in $\mathrm{Hom}(F,G)$ in order to conclude $C^*U(F,G)$ is compact. If $e_1,\ldots,e_m$ are the minimal central projections of F, then it is easy to see that $\phi \in \mathrm{Hom}(F,G)$ is injective if and only if $\phi(e_1),\phi(e_2),\ldots,\phi(e_m)$ are all non-zero and hence if and only if $\|\phi(e_i)\| = 1$, $1 \le i \le m$. Thus $C^*U(F,G)$ is indeed a compact Hausdorff space, which is open in $\mathrm{Hom}(F,G)$. We can write $G = \prod_{i=1}^{n} M(r_i)$ and note that if $P_0 = \prod_{i=1}^{n} \mathrm{Aut}(M(r_i))$, then $P_0 \subset \mathrm{Aug}(G)$. To show $C^*U(F,G)/\mathrm{Aut}(G)$ is a finite discrete space, it suffices to show $C^*U(F,G)/P_0$ is a finite discrete space. It suffices to show $\mathrm{Hom}(F,G)/P_0$ is finite. But $\mathrm{Hom}(F,G) = \prod_{i=1}^{n} \mathrm{Hom}(F,M(r_i))$. hence $\mathrm{Hom}(F,G)/P_0 = \prod_{i=1}^{n} Q(r_i)$, where for each r, $Q(r) = \mathrm{Hom}(F,M(r))/\mathrm{Aut}(M(r))$. But then $Q(r)$ is just the set of equivalence classes of r-dimensional representations of F so is finite, for each positive integer r. As $\mathrm{Aut}(G)$ is a compact Lie group, it follows that each $\mathrm{Aut}(G)$-orbit in $\mathrm{Hom}(F,G)$ is a compact manifold, in fact a homogeneous space, and since the orbit space is a compact discrete space, $\mathrm{Hom}(F,G)$ is a disjoint union of compact homogeneous spaces and therefore is a manifold.

PROPOSITION 3.11. *Let* ξ *and* ζ *be homogeneous* C^**-bundles over* X *with finite dimensional fibres. If* ϕ *is section of* $C^*U(\xi,\zeta)$*, then* ϕ *is locally trivial.*

Proof. Let $x_0 \in X$ and begin by choosing an open neighborhood W of x_0 with $\xi|W$ and $\zeta|W$ trivial. Let θ_1 and ψ_1 be trivializations of $\xi|W$ and $\zeta|W$ respectively, and let F and G be the fibres of ξ and ζ respectively. Then $C^*U(\xi,\zeta)$ is trivial over W and $\psi_1(\phi|W)\theta_1^{-1} : \varepsilon(W,F) < \varepsilon(W,G)$ is a section of the trivial bundle with fibre $C^*U(F,G)$ whose principal part we denote $f : W \to C^*U(F,G)$. Let $\phi_0 = f(x_0) : F < G$. Let K_0 denote the orbit of ϕ_0 under the action of $\mathrm{Aut}(G)$. Since the orbit space is discrete, K_0 must be compact and open in $C^*U(F,G)$. Let $V_1 = f^{-1}(K_0)$. Then V_1 is an open neighborhood of x_0 and f restricts to a map $V_1 \to K_0$. As $\mathrm{Aut}(G)$ is a compact Lie group acting transitively on K_0, the map $\psi \to \psi\phi_0$ of $\mathrm{Aut}(G)$ on K_0 is the projection, p, of a fibre bundle over K_0. We can now choose K_1 open in K_0 with $\phi_0 \in K_1$, and a local cross-section $s : K_1 \to \mathrm{Aut}(G)$ of the fibre bundle $p : \mathrm{Aut}(G) \to K_0$. It follows that for ϕ in K_1 we have $s(\phi)\phi_0 = \phi$. Finally, let $V = f^{-1}(K_1)$. Then V is an open neighborhood of x_0, so without loss of generality, we may assume $V = W$. The map $h = s \circ f : V \to \mathrm{Aut}(G)$ defines an automorphism λ of $\varepsilon(V,G)$. It is easy to see that $\lambda \circ \varepsilon(V,\phi_0) = \psi_1 \circ (\phi|V) \circ \theta_1^{-1}$, and hence if $\theta = \theta_1$ and $\psi = \lambda^{-1} \circ \psi_1$, then (θ,ψ) is a trivialization of $\phi|V$.

Our result 3.11 will be used in section seven to prove a homotopy invariance for second order C*-bundles having only finite dimensional fibres.

4. TENSOR PRODUCTS OF C*-BUNDLES

In the classification theory of Hilbert bundles, a very important role is played by the direct sum and orthogonal complement operations (on bundles) in the formation of local decompositions. For SC_1^*-bundles, the operation analogous to the direct sum of Hilbert bundles is the tensor product operation for C*-bundles. Unfortunately, for general C*-bundles the tensor product does not always seem to be defined. Tensor products of Hilbert bundles are easily formed as first shown in [9]. Tensor products of the C*-bundles arising from continuous trace C*-algebras have been defined via the local Hilbert bundle structure from which the C*-bundle structure locally arises [7]. This method, in addition to being aesthetically unpleasant, has the drawback of not being general enough for our purposes.

For the definition and basic properties for tensor products of C*-algebras we refer to [26]. In particular, if ϕ and ψ are *-homomorphisms, there is a *-homomorphism $\phi \otimes \psi$ which is injective if ϕ and ψ are. Thus the tensor product of a pair of *-isomorphisms is again a *-isomorphism.

Suppose now that ξ and ζ are C*-bundles over the space X. For each $x \in X$, form the C*-algebra $\xi(x) \otimes \zeta(x)$ and define A to be the (algebraic) *-algebra product,

$$A = \prod_{x \in X} \xi(x) \otimes \zeta(x).$$

For $s \in X(\xi)$ and $t \in X(\zeta)$, define $s \otimes t \in A$ by setting $(s \otimes t)(x) = s(x) \otimes t(x)$, $x \in X$. Let S be a fixed *total* *-subalgebra of $\Gamma(\xi)$ and T a fixed *total* *-subalgebra of $\Gamma(\zeta)$, so for each $x \in X$, $S(x)$ and $T(x)$ are dense *-subalgebras of $\xi(x)$ and $\zeta(x)$, respectively. Let $\Gamma_0 = \{s \otimes t : s \in S, t \in T\}$, and let $\Gamma_1 = \{s \otimes t : s \in \Gamma(\xi), t \in \Gamma(\zeta)\}$. Let A_i be the *-subalgebra of A generated by Γ_i, $i = 0$ or 1. Since for $i = 0$ or 1, $\Gamma_i^2 = \Gamma_i$

and $\Gamma_i^* = \Gamma_i$, it follows that A_i is the set of finite sums of members of Γ_i, and since $\Gamma_0 \subset \Gamma_1$, we have $A_0 \subset A_1 \subset A$.

REMARKS. (1) *For each* $x \in X$, $A_0(x)$ *and* $A_1(x)$ *are dense *-subalgebras of* $\xi(x) \otimes \zeta(x)$.

(2) *By* [15, *prop.* 1], *there can be at most one* C*-*bundle* ν *over* X *satisfying*

(a) *for each* $x \in X$, $\nu(x) = \xi(x) \otimes \zeta(x)$, *and*

(b) $A_0 \subset \Gamma(\nu)$.

(3) *In fact* (2) *above would still hold if we replace condition* (b) *by any one of the equivalent conditions*

(b') $\Gamma_0 \subset \Gamma(\nu)$,

(b_1) $\Gamma_1 \subset \Gamma(\nu)$,

(b_1') $A_1 \subset \Gamma(\nu)$.

(4) *By* [15, *prop.* 1], *for the existence of the* C*-*bundle* ν *in* (2) *satisfying* (a) *and* (b) *it is necessary and sufficient that either one of the following conditions holds:*

(nc) *for each* $a \in A_0$, $|a| : X \to \mathbb{R}$ *is continuous*,

(nc') *for each* $a \in A_1$, $|a| : X \to \mathbb{R}$ *is continuous*.

(5) *If* F_1 *and* F_2 *are* C*-*algebras, if* ξ *and* ζ *are the product bundles with fibres* F_1 *and* F_2 *respectively, then with* $\nu = \varepsilon(X, F_1 \otimes F_2)$, *the product* C*-*bundle over* X *with fibre* $F_1 \otimes F_2$, (2) *is true*. Indeed, the map $(a,b) \to a \otimes b$ is a bounded bilinear map $F_1 \times F_2 \to F_1 \otimes F_2$, hence is continuous, and sections can be identified with their principal parts as merely continuous vector functions on X so that (3) applies.

The following definition for the *tensor product* of ξ and ζ seems natural in view of these remarks.

DEFINITION 4.1. *We say* $\xi \otimes \zeta$ *is defined if there is a* C*-*bundle* ν *over* X *satisfying* (a) *and* (b) *of* (2), *and in this case in view of* (2) *we define* $\xi \otimes \zeta = \nu$.

By definition 4.1 and remark (5), we see that $\varepsilon(X,F_1) \otimes \varepsilon(X,F_2)$ is always defined and $\varepsilon(X,F_1) \otimes \varepsilon(X,F_2) = \varepsilon(X, F_1 \otimes F_2)$, for any C*-algebras F_1 and F_2. It is also obvious from 1.1 and 1.2 that if ξ' and ζ' are C*-bundles over X with $\alpha : \xi \cong \xi'$, $\beta : \zeta \cong \zeta'$ and if $\xi' \otimes \zeta'$ is defined, then $\xi \otimes \zeta$ is defined, and there is

$\alpha \otimes \beta : \xi \otimes \zeta \cong \xi' \otimes \zeta'$, unique such that for each $x \in X$, $(\alpha \otimes \beta)(x) = \alpha(x) \otimes \beta(x)$.

PROPOSITION 4.2. *Suppose that* ξ' *and* ζ' *are* C*-*bundles over* X' *and that* $\phi : \xi \to \xi'$, $\psi : \zeta \to \zeta'$ *are both strong* C*-*bundle maps covering* $f : X \to X'$. *If* $\xi' \otimes \zeta'$ *is defined then* $\xi \otimes \zeta$ *is defined.*

Proof. With $f^*(\Gamma(\xi')) = S$ and $f^*(\Gamma(\zeta')) = T$ denoting the images of the maps $f^* : \Gamma(\xi') \to \Gamma(f^*(\xi'))$ and $f^* : \Gamma(\zeta') \to \Gamma(f^*(\zeta'))$, respectively, we get S a full *-algebra for ξ and T a full *-algebra for ζ. If $s_1,\ldots,s_n \in S$ and $t_1,\ldots,t_n \in T$, set $s_i = f^*(s_i')$, $t_i = f^*(t_i')$ with $s_i' \in \Gamma(\xi')$ and $t_i' \in \Gamma(\zeta')$, $1 \le i \le n$. Set $a = \sum_{i=1}^{n} s_i \otimes t_i$ and $a_i' = \sum_{i=1}^{n} s_i' \otimes t_i'$. Let $\phi' : f^*(\xi') \to \xi'$ and $\psi' : f^*(\zeta') \to \zeta'$ denote the bundle maps induced by f, so both are strong C*-bundle maps. In particular, they are norm preserving on fibres. For $x \in X$, $1 \le i \le n$, $\phi'(s_i(x)) = s_i'(f(x))$ and $\psi'(t_i(x)) = t_i'(f(x))$, hence we obviously have $|a| = |a'| \circ f$. By remark (4), $|a'|$ is continuous, hence (nc) of (4) must hold for $f^*(\xi')$ and $f^*(\zeta')$ so $f^*(\xi') \otimes f^*(\zeta')$ is defined. An application of 1.1 shows that $\xi \cong f^*(\xi')$ and $\zeta \cong f^*(\zeta')$, completing the proof.

PROPOSITION 4.3. *Suppose that* ξ' *and* ζ' *are* C*-*bundles over* X' *and that* $\phi : \xi \to \xi'$, $\psi : \zeta \to \zeta'$ *are both* C*-*bundle maps covering* $f : X \to X'$. *Suppose that* $\xi \otimes \zeta$ *and* $\xi' \otimes \zeta'$ *are both defined. Then there is a unique* C*-*bundle map* $\phi \otimes \psi : \xi \otimes \zeta \to \xi' \otimes \zeta'$ *covering* f, *unique such that for each* $x \in X$, $(\phi \otimes \psi)(x) = \phi(x) \otimes \psi(x)$. *If* ϕ *and* ψ *are strong, then* $\phi \otimes \psi$ *is strong.*

Proof. There is an obvious pre-C*-bundle map $\phi \otimes \psi : \xi \otimes \zeta \to \xi' \otimes \zeta'$ covering f and unique such that for each $x \in X$, $(\phi \otimes \psi)(x) = \phi(x) \otimes \psi(x)$. It remains to show that $\phi \otimes \psi$ is an actual C*-bundle map. Since $\phi \otimes \psi$ is a *-homomorphism on each fibre, condition (1) of 1.2 is automatic. We also notice that Γ_0 is total for $\xi \otimes \zeta$ because for each $x \in X$, $A_0(x)$ is a dense *-subalgebra of $\xi(x) \otimes \zeta(x) = (\xi \otimes \zeta)(x)$ which is the linear span of $\Gamma_0(x)$. If $s \in S$ and $t \in T$, then we must show $(\phi \otimes \psi) \circ (s \otimes t)$ is continuous. Suppose first that $X = X'$ and f is the identity map. Then $(\phi \circ s) \in \Gamma(\xi')$, $\psi \circ t \in \Gamma(\zeta')$ and thus by condition (b_1) of remark (3), $(\phi \otimes \psi) \circ (s \otimes t) = (\phi \circ s) \otimes (\psi \circ t) \in \Gamma(\xi' \otimes \zeta')$ and hence by 1.2,

$\phi \otimes \psi$ is a C*-bundle map. On the other hand, consider the case of an arbitrary map $f : X \to X'$, but where ϕ and ψ are both *strong* C*-bundle maps. By 1.1, in this case, we can consider $\xi = f^*(\xi')$, $\zeta = f^*(\zeta')$ and suppose that ϕ and ψ are the C*-bundle maps induced by f. Then as in the proof of 4.2, we take $S = f^*(\Gamma(\xi'))$, $T = f^*(\Gamma(\zeta'))$, so if $s \in S$ and $t \in T$, there is $s' \in \Gamma(\xi')$ and $t' \in \Gamma(\zeta')$ with $\phi \circ s = s' \circ f$ and $\psi \circ t = t' \circ f$. Thus

$$(\phi \otimes \psi) \circ (s \otimes t) = (s' \otimes t') \circ f$$

is a continuous map of X to the total space of $\xi' \otimes \zeta'$. Again, applying 1.2, we see that $\phi \otimes \psi$ is a C*-bundle map. Consider now the general case and let $\phi' : f^*(\xi') \to \xi'$ and $\psi' : f^*(\zeta') \to \zeta'$ be the induced C*-bundle maps. Applying 1.1, there is $\theta_1 : \xi \to f^*(\xi')$ and $\theta_2 : \zeta \to f^*(\zeta')$ over X, unique C*-bundle maps such that $\phi = \phi' \circ \theta_1$ and $\psi = \psi' \circ \theta_2$. We then have $\phi \otimes \psi = (\phi' \otimes \psi') \circ (\theta_2 \otimes \theta_1)$ expressing $\phi \otimes \psi$ as a composition of C*-bundle maps and showing that $\phi \otimes \psi$ is itself a C*-bundle map. If ϕ and ψ are strong, then for $x \in X$, $\phi(x)$ and $\psi(x)$ are *-isomorphisms so $\phi(x) \otimes \psi(x) = (\phi \otimes \psi)(x)$ is also a *-isomorphism, hence $\phi \otimes \psi$ is strong and the proof is complete.

If (Y,d) is a metric space, $y \in Y$, $V \subset Y$, then $d(y,V)$, the distance from y to V, is the infimum of the set $\{d(y,s) : s \in V\}$. We say the C*-algebra B is *approximately finite dimensional* if for each finite subset $F \subset B$ and each $\varepsilon > 0$ there is a finite dimensional C*-subalgebra B_0 of B such that $d(y,B_0) < \varepsilon$ for each $y \in F$. It is known that a separable C*-algebra is approximately finite dimensional if and only if it is a direct limit of finite dimensional C*-algebras.

It is obvious from remark (4) that if there is an open cover $\mathcal{U}$ of X such that for each $U \in \mathcal{U}$, $(\xi|U) \otimes (\zeta|U)$ is defined, then $\xi \otimes \zeta$ is defined.

DEFINITION 4.4. *We say* ξ *is approximately locally trivial if* ξ *satisfies the condition*

(a.l.t.) *if* $x_0 \in X$ *and* $\varepsilon > 0$, *and if* K *is a finite subset of* $\xi(x_0)$, *then there is* U *an open neighborhood of* x_0 *and* ξ_0 *a trivial* C*-*subbundle of* $\xi|U$ *such that for each* $a \in K$,

$$d(a,\xi_0(x_0)) < \varepsilon.$$

It is easy to see from the results of the preceeding section, that if all the fibres of ξ are approximately finite dimensional, then ξ is approximately locally trivial. Of course, if ξ is locally trivial then ξ is approximately locally trivial. This means that the class of all approximately locally trivial C*-bundles is fairly large. In particular, it includes all the C*-bundles arising in the representation theory of CCR-algebras, and any locally trivial bundles one might be inclined to consider.

THEOREM 4.5. *If* ξ *and* ζ *are approximately locally trivial, then* $\xi \otimes \zeta$ *is defined.*

Proof. We shall show that for each $a \in A_0$, the function $|a| : X \to \mathbb{R}$ is continuous so that remark (4) can be applied. Set $f = |a|$. To show f is continuous at $x_0 \in X$, it suffices to show that for each $\delta > 0$ there is an open neighborhood U of x_0 and a continuous function $g : U \to \mathbb{R}$ such that for every $x \in U$, $|g(x) - f(x)| < \delta$. Since A_0 consists of finite sums of members of Γ_0, there are sections $s_1,\ldots,s_n \in S$ and $t_1,\ldots,t_n \in T$ such that

$$a = \sum_{i=1}^{n} s_i \otimes t_i .$$

Fix $x_0 \in X$. Since we need only work locally around x_0, the continuity of the norm guarantees that there is no loss of generality in assuming all the sections $s_1,\ldots,s_n$, $t_1,\ldots,t_n$ are bounded, say by a real number $M \geq 1$. Suppose now that $\delta > 0$ is given. Define $\varepsilon = [2n(M+\delta)]^{-1}\delta$. Since ξ and ζ are approximately locally trivial, there is an open neighborhood U of x_0 and C*-subbundles ξ_0 and ζ_0 of $\xi|U$ and $\zeta|U$ respectively such that ξ_0 and ζ_0 are trivial and for each i, with $1 \leq i \leq n$, $d(s_i(x_0),\xi_0(x_0)) < \varepsilon$ and $d(t_i(x_0),\zeta(x_0)) < \varepsilon$. For each i, $1 \leq i \leq n$, choose $s_i' \in \Gamma(\xi_0)$ and $t_i' \in \Gamma(\zeta_0)$ such that $\|s_i(x_0) - s_i'(x_0)\| < \varepsilon$ and $\|t_i(x_0) - t_i'(x_0)\| < \varepsilon$. Since the norm is continuous on total spaces, on making U smaller if necessary, we may assume that $\|s_i' - (s_i|U)\| < \varepsilon$ and $\|t_i' - (t_i|U)\| < \varepsilon$. In particular, as $\varepsilon < \delta$, the sections $s_1',\ldots,s_n'$, $t_1',\ldots,t_n'$ are all bounded by $M+\delta$. Define

$$a' = \sum_{i=1}^{n} s_i' \otimes t_i' ,$$

so $a' \in \Gamma_b(\xi_0 \otimes \zeta_0)$, $\xi_0 \otimes \zeta_0$ being defined because ξ_0 and ζ_0 are trivial. Setting $g = |a'| : U \to \mathbb{R}$, g is continuous by remark (4) and it is routine to see that for $x \in U$, $|g(x) - f(x)| < \delta$, to complete the proof.

Let ξ be any C*-bundle over X and suppose ξ_1 and ξ_2 are subbundles of ξ. We say ξ_1 and ξ_2 *commute* if for each $x \in X$, all elements of $\xi_1(x)$ commute with all elements of $\xi_2(x)$.

PROPOSITION 4.6. *Let* ξ *be a* C*-*bundle over* X *and suppose* ξ_1 *and* ξ_2 *are* C*-*subbundles of* ξ *which commute. Suppose that all fibres of* ξ_1 *and* ξ_2 *are finite dimensional. Then there is* $m : \xi_1 \otimes \xi_2 \to \xi$ *a unique* C*-*bundle map such that for each* $x \in X$, $m(x) : \xi_1(x) \otimes \xi_2(x) \to \xi(x)$ *is the map defined by the multiplication of* $\xi(x)$.

Proof. There is obviously a pre-C*-bundle map $m : \xi_1 \otimes \xi_2 \to \xi$ unique such that for each $x \in X$, $m(x) : \xi_1(x) \otimes \xi_2(x) \to \xi(x)$ is the map defined by the multiplication of $\xi(x)$. The fact that $m(x)$ is a *-homomorphism follows easily from the fact that ξ_1 and ξ_2 commute. It only remains to show that m is an actual C*-bundle map. If $s_1 \in \Gamma(\xi_1)$ and $s_2 \in \Gamma(\xi_2)$, then $m \circ (s_1 \otimes s_2) = s_1 s_2 \in \Gamma(\xi)$, hence as $\{s_1 \otimes s_2 : s_1 \in \Gamma(\xi_1), s_2 \in \Gamma(\xi_2)\}$ is total for $\xi_1 \otimes \xi_2$, an application of 1.2 completes the proof.

The tensor product we have defined does not correspond to the tensor product of C*-algebras via Γ_b. If ξ and ζ are approximately locally trivial C*-bundles over X and Y respectively, we can define their *exterior tensor product* $\xi \check{\otimes} \zeta = \pi_1^*(\xi) \otimes \pi_2^*(\zeta)$ where $\pi_1 : X \times Y \to X$ and $\pi_2 : X \times Y \to Y$ are the projection maps. If $a \in \Gamma_b(\xi)$, $b \in \Gamma_b(\zeta)$ we define $a \check{\otimes} b = \pi_1^*(a) \otimes \pi_2^*(b) \in \Gamma_b(\xi \check{\otimes} \zeta)$.

PROPOSITION 4.7. *There is an injective *-homomorphism* $\theta : \Gamma_b(\xi) \otimes \Gamma_b(\zeta) \cong \Gamma_b(\xi \check{\otimes} \zeta)$ *unique such that if* $a \in \Gamma_b(\xi)$ *and* $b \in \Gamma_b(\zeta)$, *then* $\theta(a \otimes b) = a \check{\otimes} b$. *If* X *and* Y *are compact Hausdorff, then* θ *is a *-isomorphism (onto).*

Proof. Let A be the algebraic tensor product of $\Gamma_b(\xi)$ and $\Gamma_b(\zeta)$ which is then a dense *-subalgebra of $\Gamma_b(\xi) \otimes \Gamma_b(\zeta)$. Then there is $\theta_0 : A \to \Gamma_b(\xi \check{\otimes} \zeta)$ a *-homomorphism unique such that $\theta_0(a \otimes b) = a \check{\otimes} b$, $a \in \Gamma_b(\xi)$, $b \in \Gamma_b(\zeta)$. By [26, 1.22.7], we have $\|s\| \leq \|\theta_0(s)\|$ for each $s \in A$. On the other hand if $(x,y) \in X \times Y$ and if $e_x : \Gamma_b(\xi) \to \xi(x)$ and $e_y : \Gamma_b(\zeta) \to \zeta(y)$ are the evaluation maps, then e_x and e_y are *-homomorphisms and $e_x \otimes e_y :$ $\Gamma_b(\xi) \otimes \Gamma_b(\zeta) \to \xi(x) \otimes \zeta(y)$ is a *-homomorphism. Hence $\|s(x,y)\| \leq \|s\|$ for any $s \in A$ and therefore $\|\theta_0(s)\| \leq \|s\|$. Thus θ_0 is isometric. If X and Y are compact Hausdorff, then as $C(X \times Y) = C(X) \otimes C(Y)$, [18, 1.7] applies to complete the proof.

We can relate our tensor product of C*-bundles directly to C*-algebras in another useful way. If A and B are C*-algebras which are algebras over the same abelian C*-algebra C, so that $\|fa\| \leq \|f\| \|a\|$, $f \in C$, $a \in A$ or B and with $(fa)^* = f^*a^*$, then we define

$$A \otimes_C B = A \otimes B/I$$

where I is the smallest two-sided closed ideal of $A \otimes B$ containing all elements of the form $(fa) \otimes b - a \otimes (fb)$, $f \in C$, $a \in A$, $b \in B$. We denote $a \otimes_C b = a \otimes b + I$, for $a \in A$, $b \in B$. If ξ and ζ are C*-bundles over X we define

$$\Gamma_b(\xi) \otimes_X \Gamma_b(\zeta) = \Gamma_b(\xi) \otimes_C \Gamma_b(\zeta) \quad \text{where} \quad C = C_b(X),$$

and we denote $a \otimes_X b = a \otimes_C b$, $a \in \Gamma_b(\xi)$, $b \in \Gamma_b(\zeta)$.

Let $d : X \to X \times X$ be the diagonal map. Then $d^*(\xi \check{\otimes} \zeta) = \xi \otimes \zeta$ as $\pi_i \circ d = id_X$, $i = 1,2$, where $\pi_1, \pi_2 : X \times X \to X$ are the coordinate projections. Thus $d^* : \Gamma_b(\xi \check{\otimes} \zeta) \to \Gamma_b(\xi \otimes \zeta)$ is a *-homomorphism, which is surjective if $X \times X$ is paracompact Hausdorff. With $\theta : \Gamma_b(\xi) \otimes \Gamma_b(\zeta) \to \Gamma_b(\xi \check{\otimes} \zeta)$ the *-homomorphism of 4.7, it is obvious that $\theta(I) \subset \operatorname{Ker} d^*$, and hence there is a *-homomorphism

$$\overline{\theta} : \Gamma_b(\xi) \otimes_X \Gamma_b(\zeta) \to \Gamma_b(\xi \otimes \zeta)$$

unique so that $\overline{\theta}(s + I) = d^*(\theta(s))$, for $s \in \Gamma_b(\xi) \otimes \Gamma_b(\zeta)$. Clearly then $\overline{\theta}$ is also unique such that if $a \in \Gamma_b(\xi)$, $b \in \Gamma_b(\zeta)$, then $\overline{\theta}(a \otimes_X b) = a \otimes b \in \Gamma_b(\xi \otimes \zeta)$.

PROPOSITION 4.8. *If* X *is a compact Hausdorff space, then* $\overline{\theta} : \Gamma(\xi) \otimes_X \Gamma(\zeta) \to \Gamma(\xi \otimes \zeta)$ *is a *-isomorphism which is unique such* *that* $\overline{\theta}(a \otimes_X b) = a \otimes b$ *for* $a \in \Gamma(\xi)$, $b \in \Gamma(\zeta)$.

Proof. If we identify $\Gamma(\xi) \otimes \Gamma(\zeta) = \Gamma(\xi \check{\otimes} \zeta)$ via the *-isomorphism θ of 4.7, then we need only show that $I = \text{Ker } d^*$. Let $\eta \subset \xi \check{\otimes} \zeta$ be the C*-subbundle such that

$$\eta(x,y) = \begin{cases} 0 \ , \ x = y \\ \xi(x) \otimes \zeta(y), \ x \neq y \end{cases}$$

for $(x,y) \in X \times X$. By proposition 1.4, $\text{Ker } d^* = \Gamma(\eta)$. By [8, 10.4.2], there is a C*-subbundle $\beta \subset \xi \check{\otimes} \zeta$ with $I = \Gamma(\beta)$. Since $I \subset \text{Ker } d^*$, we have $\beta \subset \eta$, hence $\beta(x,x) = 0$ if $x \in X$. If $x,y \in X$, $x \neq y$, we can choose $f : X \to [0,1]$ with $f(x) = 1$ and $f(y) = 0$. If $a \in \xi(x)$, $b \in \zeta(y)$, then there is $s \in \Gamma_b(\xi)$ with $s(x) = a$ and there is $t \in \Gamma_b(\zeta)$ with $t(y) = b$. Then $v = (fs) \check{\otimes} t - s \check{\otimes} (ft)$ belongs to I and therefore $v(x,y) \in \beta(x,y)$. But,

$$v(x,y) = f(x)s(x) \otimes t(y) - s(x) \otimes f(y)t(y) = a \otimes b.$$

Thus $a \otimes b \in \beta(x,y)$ for any $a \in \xi(x)$ and any $b \in \zeta(y)$. It follows that $\beta(x,y) = \xi(x) \otimes \zeta(y) = \eta(x,y)$ and thus $\beta = \eta$. Consequently, $I = \text{Ker } d^*$ and the proof is complete.

We remark that if $\infty \in X$ and $\xi(\infty) = 0 = \zeta(\infty)$ (see the remarks following 1.4), then the ideal I in 4.8 is also generated by elements of the form $(fa) \otimes b - a \otimes (fb)$ with $f \in C(X)$ such that $f(\infty) = 0$, $a \in \Gamma_b(\xi)$, $b \in \Gamma_b(\zeta)$. It follows that if X is only locally compact we can replace Γ_b by Γ_0 in 4.8 and replace $C_b(X)$ by $C_0(X)$ to obtain

$$\Gamma_0(\xi) \otimes_{C_0(X)} \Gamma_0(\zeta) = \Gamma_0(\xi \otimes \zeta).$$

5. BUNDLE COMMUTANTS

As we shall replace the direct sum operation of Hilbert bundle theory by the tensor product operation for C*-bundles, we need an analogue of the *orthogonal complement*. That analogue for SC_1^*-bundles is the *commutant*.

DEFINITION 5.1. Let ξ be a C*-bundle over X and let ζ be a subbundle of ξ over X. They by the *commutant of* ζ *in* ξ, denoted ζ', we mean the unique subbundle of ξ defined by the requirement that for each $x \in X$, $\zeta'(x)$ is the ordinary commutant of $\zeta(x)$ in $\xi(x)$, so

$$\zeta'(x) = \{a \in \xi(x) : \text{for each } b \in \zeta(x),\ ab = ba\}.$$

The problem here is that when dealing with C*-bundles we would like for commutants of C*-subbundles to again be C*-subbundles. This is too much to ask. Even with Hilbert bundles, the orthogonal complement of a Hilbert subbundle is not necessarily again a Hilbert subbundle. Counterexamples are everywhere and easy to construct because of [13, prop. 1.6]. The analogous problem obviously occurs for C*-bundles also. But, the analogue of [9, prop. 17 , pg.247] is true, the proof of which is the main purpose of this section.

If ξ is a C_1^*-bundle over X and ζ is a C*-subbundle of ξ over X we say ζ is a C_1^*-*subbundle* of ξ over X to mean that for each $x \in X$ the identity of $\xi(x)$ belongs to $\zeta(x)$, so in particular, ζ is itself a C_1^*-bundle. Notice that if ξ is a simple C_1^*-bundle then each fibre is finite dimensional, since each fibre is simple dual and has an identity. Thus, for each $x \in X$ there is a positive integer $n(x)$ with $\xi(x) = M(n(x))$, the C*-algebra of all complex $n(x) \times n(x)$ matrices. We must therefore consider simple C_1^*-subalgebras of $M(n)$ where n is a positive integer.

Let A be a simple C_1^*-subalgebra of $M(n)$, so A contains the identity of $M(n)$, and for some positive integer m, $A \cong M(m)$. Choose a system $S = (e_{ij})$ of matrix units for A, so we have

$$(1) \qquad e_{\alpha\beta}e_{\gamma\delta} = \begin{cases} 0 & \beta \neq \gamma \\ e_{\alpha\delta}, & \beta = \gamma \end{cases}, \quad 1 \le \alpha,\beta,\gamma,\delta \le m$$

$$(2) \qquad e_{ij}^* = e_{ji}, \quad 1 \le i,j \le m.$$

Using the system S, we define a linear map $\phi_S : M(n) \to M(n)$ by the formula

$$(*) \qquad \phi_S(b) = \sum_{k=1}^{m} e_{k1} b e_{1k} .$$

Since $\Sigma_k e_{k1}e_{1k} = \Sigma_k e_{kk} = 1$, the identity of $M(n)$, if $b \in A'$, then $\phi_S(b) = b$. On the other hand, if $1 \le i,j \le m$, then for any $b \in M(n)$ we have

$$e_{ij}\phi_S(b) = e_{i1}be_{1j} = \phi_S(b)e_{ij} ,$$

and therefore $\phi_S(b)$ belongs to A', the commutant of A in $M(n)$. Thus ϕ_S is a linear retraction (algebraic projection) of $M(n)$ onto A'. It would be nice if ϕ_S were independent of the particular system S chosen, but such is not the case. However, it is easy to see that the multiplication of $M(n)$ determines a unique *-isomorphism

$$A \otimes A' \cong M(n).$$

We now examine the situation for bundles. *Let* ξ *be a simple* C_1^**-bundle over* X *and let* β *be a homogeneous simple* C_1^**-subbundle of* ξ*, fixed for the remainder of this section.*

LEMMA 5.2. *The commutant,* β'*, of* β *in* ξ *is locally full.*

Proof. Let $x_0 \in X$. We must find an open neighborhood U of x_0 such that $\beta'|U$ is *full*. Simply choose any open neighborhood of x_0 such that $\beta|U$ is trivial. This can be done by proposition 3.6. We obtain therefore a family of sections (e_{ij}) in $\Gamma_b(U,\beta)$ satisfying conditions (1) and (2) pointwise. Thus, $S = (e_{ij})$ is a system of matrix units for a C*-subalgebra of $\Gamma_b(U,\beta)$. Suppose that $x \in U$ and

$a \in \beta'(x)$. Choose any section $b \in \Gamma(U,\xi)$ with $b(x) = a$. Define $b' = \phi_S(b)$ using equation (*) on our system S. By our previous remarks concerning (*), we know that for each $y \in U$, $b'(y) \in \beta'(y)$ and that $b'(y) = \phi_{S(y)}(b(y))$, where $S(y) = (e_{ij}(y))$. In particular, $b'(x) = b(x) = a$ because $a \in \beta'(x)$ already. Thus, $b' \in \Gamma(U,\beta')$ and $b'(x) = a$, completing the proof.

It is now easy to see that β' satisfies all the requirements of being a simple C_1^*-subbundle of ξ, with two possible exceptions. We must show that the bundle projection of β' is an open map, and since all Banach bundles and C*-bundles are assumed full, we must show that β' is full. The fact that β' is locally full makes it easy to see that the bundle projection of β' is an open map. If X is completely regular, then since β' is locally full, it is full. This gives the next result.

THEOREM 5.3. *The commutant,* β', *of* β *in* ξ *is also a simple* C_1^**-subbundle of* ξ, *if* X *is completely regular.*

COROLLARY 5.4. *There is a unique* C_1^**-bundle isomorphism* $m : \beta \otimes \beta' \cong \xi$ *over* X *such that for each* $x \in X$, $m(x) : \beta(x) \otimes \beta'(x) \cong \xi(x)$ *is the map defined by the multiplication in* $\xi(x)$.

Proof. The existence of the C*-bundle map is given by proposition 4.6. Since in this case for each $x \in X$, $m(x) : \beta(x) \otimes \beta'(x) \cong \xi(x)$, m is a strong C*-bundle map covering the identity so by 1.1 or [13, prop. 1.4], m is a C_1^*-bundle isomorphism.

6. BANACH BUNDLE PASTING

As Banach bundles can be intuitively thought of as "vector bundles with varying fibre dimension", it is reasonable to try to develop a method for "pasting together" homogeneous Banach bundles having different fibres to obtain one with varying fibres. Since fibre dimensions can vary, we clearly cannot require that we have "coordinate transformations" in any ordinary sense of the term. Loosely speaking, the bundles with lower fibre dimension are on overlaps merely embedded into the bundles with larger fibre dimension, a non-symmetrical situation.

Suppose X is a space and $\mathcal{U} = (U_\alpha)_D$ is a family of open sets covering X which is indexed by D. Suppose $(\xi_\alpha)_D$ is a family of Banach bundles indexed by D so that for each $\alpha \in D$, U_α is the base space of ξ_α. If $(\alpha_1,\ldots,\alpha_n)$ is an n-tuple in D^n, $a = (\alpha_1,\ldots,\alpha_n)$ we set

$$U_a = U_{\alpha_1,\ldots,\alpha_n} = \bigcap_{i=1}^{n} U_{\alpha_i}$$

and

$$\xi_a = \xi_{\alpha_1,\ldots,\alpha_n} = \xi_{\alpha_1}|U_a .$$

We want to try to paste the family (ξ_α) together to form a Banach bundle over X. If $\alpha,\beta \in D$ and $U_{\alpha,\beta} \neq \emptyset$, then we can hope that either $\xi_{\alpha,\beta} < \xi_{\beta,\alpha}$ or $\xi_{\beta,\alpha} < \xi_{\alpha,\beta}$, but in general, we cannot ask for both, as in the classical setting. What we need is some type of order on the index set D which will always enable us to select the correct possible direction of embedding. Obviously in practice, the order must be dictated by the fibre dimensions of the various component bundles (ξ_α). Therefore, we shall assume a relation $\prec$ on the index set D which is reflexive and transitive such that

(1) if $\alpha,\beta \in D$ and $U_{\alpha,\beta} \neq \emptyset$, then $\alpha \prec \beta$ or $\beta \prec \alpha$,

(2) if $x \in X$, then the set $D(x) = \{\alpha \in D : x \in U_\alpha\}$ is *directed* under $\prec$.

We denote $\succ = \prec^{-1}$ and identify relations on D as subsets of the set $D \times D$. Condition (1) merely says that when the overlap information is needed it is there, whereas condition (2) will allow us to determine the fibres of the prospective bundle under construction. As our generalized coordinate transformations we take a family of embeddings $(\phi_{\beta,\alpha})_\succ$ for (ξ_α) indexed by $\succ$, which means for each pair $(\alpha,\beta) \in D \times D$ with $\alpha \prec \beta$, we are given $\phi_{\beta\alpha} : \xi_{\alpha,\beta} < \xi_{\beta,\alpha}$, a Banach bundle embedding. For compatibility we require

(3) if $\alpha \prec \beta$ and $\beta \prec \gamma$, then
$$(\phi_{\gamma,\beta} \mid U_{\alpha,\beta,\gamma}) \circ (\phi_{\beta,\alpha} \mid U_{\alpha,\beta,\gamma}) = \phi_{\gamma,\alpha} \mid U_{\alpha,\beta,\gamma},$$
and consequently

(4) for each $\alpha \in D$, $\phi_{\alpha\alpha}$ is the identity on ξ_α.

We will usually use the obvious "abuse" in notation that allows the equation of (3) to be written simply as $\phi_{\gamma,\beta} \circ \phi_{\beta,\alpha} = \phi_{\gamma,\alpha}$.

Notice that since these embeddings are isometric on fibres they are contractive. To simplify notation and terminology, we shall say ξ is a B_1-bundle to mean ξ is a Banach bundle and say ϕ is a B_1-bundle map to mean ϕ is a Banach bundle map which is contractive on fibres. We denote the resulting category by B_1.

REMARK. If $\alpha \prec \beta$ and $\beta \prec \alpha$, then obviously $\phi_{\beta,\alpha} : \xi_{\alpha,\beta} \cong \xi_{\beta,\alpha}$ and $\phi^{-1}_{\beta,\alpha} = \phi_{\alpha,\beta}$. Thus taking for $\prec$ the relation $D \times D$ would reduce us exactly to the ordinary classical situation, e.g. [8, 10.1.13], so we have an ordinary cocycle. Consequently we refer to conditions (3) and (4) for $(\phi_{\alpha,\beta})$ as *cocycle conditions* and if $(\phi_{\alpha,\beta})$ satisfies the cocycle conditions we shall call $(\phi_{\alpha,\beta})$ a *cocycle for* $(\xi_\alpha)_D$ *over* $(U_\alpha)_D$ *with respect to* $\prec$.

Given that $z = (\phi_{\beta,\alpha})_\prec$ is a cocycle for $(\xi_\alpha)_D$, we want to construct a Banach bundle $b(z)$ over X and a family $(\phi_\alpha)_D$ of embeddings where for each $\alpha \in D$, $\phi_\alpha : \xi_\alpha < b(z)|U_\alpha$ and if $\alpha \prec \beta$, then $\phi_\beta \circ \phi_{\beta,\alpha} = \phi_\alpha|U_{\alpha,\beta}$. If $x \in X$, then the family $(\phi_{\beta,\alpha}(x))$ is a direct family of isometric embeddings of Banach spaces, so we can form the direct limit Banach space, $E(x)$, in the category of Banach spaces and linear contractions and for each $\alpha \in D(x)$ we get an isometric embedding $\phi_{x,\alpha} : \xi_\alpha(x) < E(x)$ such that if $\alpha \prec \beta$ and $\alpha,\beta \in D(x)$, then

(5) $$\phi_{x,\beta} \circ \phi_{\beta\alpha}(x) = \phi_{x,\alpha} .$$

We define a surjective map $p : B \to X$ by requiring that for each $x \in X$, $p^{-1}(x) = \{x\} \times E(x) \cong E(x)$. We will simply identify $p^{-1}(x) = E(x)$ in the obvious fashion. If $x \in X$, and for each $\alpha \in D(x)$, $u_\alpha : \xi_\alpha(x) \to F$ is a linear contraction mapping into the Banach space F such that when $\alpha \prec \beta$, $u_\beta \circ \phi_{\beta\alpha}(x) = u_\alpha$, then the universal property of the direct limit guarantees existence of a linear contraction $u : E(x) \to F$ unique such that for every $\alpha \in D(x)$, $u \circ \phi_{x,\alpha} = u_\alpha$.

Giving B the weakest topology for which $p : B \to X$ is a continuous map defines a bundle $\mathcal{B}$ over X such that $\Gamma(\mathcal{B}) = \Pi(\mathcal{B})$ is simply the product of the fibres $p^{-1}(x)$, $x \in X$. Thus every selection of $\mathcal{B}$ is continuous. If $\alpha \in D$, $V \subset U_\alpha$, and $s \in \Gamma(V,\xi_\alpha)$, then we define $\tilde{s} \in \Pi(V,\mathcal{B})$ by setting for $x \in V$,

$$\tilde{s}(x) = \phi_{x,\alpha}(s(x)).$$

Let Γ be the set of all $s \in \Pi(\mathcal{B})$ such that for each $x \in X$ there is W open in X, $\alpha \in D$ with $x \in W \subset U_\alpha$, and $t \in \Gamma(W,\xi_\alpha)$ such that $s|W = \tilde{t}$.

We claim that Γ is a vector subspace of $\Pi(\mathcal{B})$. Since $\widetilde{\lambda t} = \lambda\tilde{t}$ for any scalar λ and local section t of any ξ_α, it is clear that Γ is closed under scalar multiplication. Suppose that $s_1, s_2 \in \Gamma$ and $x_0 \in X$. There are indices $\alpha_1, \alpha_2 \in D$ and open neighborhoods V_1 and V_2 of x_0 with $V_i \subset U_{\alpha_i}$, $i = 1,2$, and there are sections $t_i \in \Gamma(V_i,\xi_{\alpha_i})$, with $s_i|V_i = \tilde{t}_i$, $i = 1$ or 2. It follows that α_1 and α_2 belong to the directed set $D(x_0)$, so there is $\alpha \in D(x_0)$ with $\alpha_1 \prec \alpha$ and $\alpha_2 \prec \alpha$. Let $W = V_1 \cap V_2 \cap U_\alpha$, and for $i = 1$ or 2 set $r_i = \phi_{\alpha\alpha_i} \circ (t_i|W) \in \Gamma(W,\xi_\alpha)$. From equation (5) we see that $s_i|W = \tilde{r}_i$, $i = 1,2$ hence $(s_1 + s_2)|W = \tilde{r}_1 + \tilde{r}_2 = (r_1 + r_2)^\sim$ with $r_1 + r_2 \in \Gamma(W,\xi_\alpha)$. This completes the proof of the claim that Γ is a vector subspace of $\Pi(\mathcal{B})$. If $s \in \Gamma$ and $s|V = \tilde{t}$, $t \in \Gamma(V,\xi_\alpha)$, then $\|s(x)\| = \|t(x)\|$, $x \in V$, hence $|s|$ on V agrees with $|t|$. It follows that for each $s \in \Gamma$, the map $|s| : X \to \mathbb{R}$ is continuous. Since $\cup\{\phi_{x,\alpha}(\xi_\alpha(x)) : \alpha \in D(x)\}$ is dense in $E(x)$, $x \in X$, we see that $\Gamma(x)$ is dense in the Banach space $p^{-1}(x)$, if X is a completely regular space.

By [18, prop. 1.6], there is a unique topology on B for which $\xi = (p,B,X)$ is a Banach bundle (B_1-bundle) and $\Gamma \subset \Gamma(\xi)$, in case X is a completely regular space. We denote $\xi = b(z)$ to record the dependence of ξ on the cocycle $z = (\phi_{\beta\alpha})$. In the case where all bundles are C*-bundles, and the cocycle consists of C*-bundle maps, the technique used to show that Γ is closed under addition shows, on replacing addition by multiplication, that Γ is closed under multiplication, so Γ is a *-subalgebra and by [15, prop. 1] ξ is a C*-bundle. Now, for $\alpha \in D$, there is a pre-Banach bundle map $\phi_\alpha : \xi_\alpha \to \xi|U_\alpha$ over U_α unique such that for each $x \in U_\alpha$, $\phi_\alpha(x) = \phi_{x,\alpha}$. By equation (5), if $\alpha \prec \beta$, then $\phi_\beta \circ \phi_{\beta\alpha} = \phi_\alpha|U_{\alpha,\beta}$. Assuming X is completely regular, if Γ_α is the set of sections of ξ_α having support inside some closed subset of U_α, then Γ_α is full for ξ_α; and if $s \in \Gamma_\alpha$, then it is clear that $\tilde{s} \in \Gamma(U_\alpha, \xi)$. But $\tilde{s} = \phi_\alpha \circ s$, so by proposition 1.2, $\phi_\alpha : \xi_\alpha \to \xi|U_\alpha$ is a B_1-bundle map. Of course if (ξ_α) is a family of C*-bundles and if z consists of C*-bundle maps, then the maps $\phi_\alpha : \xi_\alpha \to \xi|U_\alpha$ are C*-bundle maps. In either the Banach or C*-case, each ϕ_α is an embedding since it is isometric on fibres, so $\phi_\alpha : \xi_\alpha < \xi|U_\alpha$, $\alpha \in D$.

Let H_1 denote the full subcategory of B_1 whose objects are Hilbert bundles. *For the remainder of this section, let* $\mathcal{C}$ *denote any of the symbols* B_1, C^*, H_1, *or* C^*_1. *We say* $(\phi_{\beta\alpha})$ *is a* $\mathcal{C}$-*cocycle for* (ξ_α) *to mean* $(\phi_{\beta\alpha})$ *is a cocycle such that for any* $\alpha \prec \beta$, $\phi_{\beta\alpha}$ *is a* $\mathcal{C}$-*bundle map of* $\mathcal{C}$-*bundles*, with ξ_α a $\mathcal{C}$-bundle.

Our preceeding discussion hopefully makes the next theorem appear quite natural.

THEOREM 6.1. *Let* X *be a completely regular space and* $z = (\phi_{\beta\alpha})$ *a* $\mathcal{C}$-*cocycle for* (ξ_α) *over* $(U_\alpha) = \mathcal{U}$ *with respect to* $\prec$, *a reflexive and transitive relation on* D *satisfying* (1) *and* (2). *Then there is a* $\mathcal{C}$-*bundle* $\xi = b(z)$ *over* X *and a family of* $\mathcal{C}$-*bundle maps* (ϕ_α) *such that for* $\alpha \in D$, $\phi_\alpha : \xi_\alpha < \xi|U_\alpha$ *and satisfying*

(a) $\xi = b(z)$ *is unique up to isomorphism over* X *with* $\phi_\beta \circ \phi_{\beta\alpha} = \phi_\alpha|U_{\alpha,\beta}$, $\alpha \prec \beta$, *and* (b) *or* (c) *below,*

(b) *if* ζ *is a* $\mathcal{C}$-*bundle over* X *and if for each* $\alpha \in D$, $\psi_\alpha : \xi_\alpha \to \zeta|U_\alpha$ *is a* $\mathcal{C}$-*bundle map over* U_α *such that* $\psi_\beta \circ \phi_{\beta\alpha} = \psi_\alpha$ *for* $\alpha \prec \beta$, *then there is a* $\mathcal{C}$-*bundle map* $\psi : \xi \to \zeta$ *over* X *unique such that for every* $\alpha \in D$, $\psi \circ \phi_\alpha = \psi_\alpha$;

(c) *for each* $x \in X$, $\cup\{\phi_\alpha(\xi_\alpha(x)) : \alpha \in D(x)\}$ *is dense in* $\xi(x)$;

(d) *if in* (b), *for each* $\alpha \in D$, $\psi_\alpha : \xi_\alpha < \zeta$, *then* $\psi : \xi < \zeta$, *and if in addition for each* $x \in X$, $\cup\{\psi_\alpha(\xi_\alpha(x)) : \alpha \in D(x)\}$ *is dense in* $\zeta(x)$, *then* $\psi : \xi \cong \zeta$.

Proof. It is obvious that (b) or (c) implies uniqueness in (a). Our preceeding discussion established existence in (a) for the cases $\mathcal{C} = B_1$ or $\mathcal{C} = C^*$, and the case $\mathcal{C} = C_1^*$ is obvious from the case $\mathcal{C} = C^*$. Given (b), then (c) and (d) are clear from our construction of ξ. Thus we need only prove (b) since the case $\mathcal{C} = H_1$ is obvious from $\mathcal{C} = B_1$. The universal property of the direct limit applied in each fibre of ξ gives a pre-$\mathcal{C}$-bundle map $\psi : \xi \to \zeta$ over X unique such that for each $\alpha \in D$, $\psi \circ \phi_\alpha = \psi_\alpha$. With Γ defined as in our preceeding discussion, if $s \in \Gamma$ and $s|V = \tilde{t}$, $t \in \Gamma(V,\xi_\alpha)$, then

$$\psi \circ (s|V) = \psi \circ \phi_\alpha \circ t = \psi_\alpha \circ t$$

is continuous, hence $\psi \circ s$ is continuous so an application of proposition 1.2 completes the proof.

COROLLARY 6.2. *Let* X *be a completely regular space,* A *a closed subset of* X *and* U *an open neighborhood of* A *in* X. *Let* ξ *be a* $\mathcal{C}$*-bundle over* U *and* ζ *a* $\mathcal{C}$*-bundle over* $X\backslash A$. *Let* $\phi : \xi|U\backslash A < \zeta|U\backslash A$ *be a* $\mathcal{C}$*-bundle embedding over* $U\backslash A$. *Then*

(a) *there is a* $\mathcal{C}$*-bundle* $\xi \cup_\phi \zeta$ *over* X *and* $\mathcal{C}$*-bundle embeddings* $\theta : \xi < (\xi \cup_\phi \zeta)|U$, $\psi : \zeta \cong (\xi \cup_\phi \zeta)|X\backslash A$ *with* $\xi \cup_\phi \zeta$ *unique up to isomorphism over* X *so that* $\psi^{-1} \circ \theta = \phi$, $\theta : \xi|A \cong (\xi \cup_\phi \zeta)|A$, and

(b) *if* γ *is a* $\mathcal{C}$*-bundle over* X, *if* $\bar{\theta} : \xi \to \gamma|U$ *and* $\bar{\psi} : \zeta \to \gamma|X\backslash A$ *are* $\mathcal{C}$*-bundle maps over* U *and* $X\backslash A$ *respectively such that* $\bar{\psi} \circ \phi = \bar{\theta}|U\backslash A$, *then there is a unique* $\mathcal{C}$*-bundle map* $\lambda : \xi \cup_\phi \zeta \to \gamma$ *over* X *such that* $\lambda \circ \theta = \bar{\theta}$ *and* $\lambda \circ \psi = \bar{\psi}$.

COROLLARY 6.3. *Let* X *be a completely regular space,* U *an open subset of* X, *and* ε *the trivial* C_1^**-bundle over* X *with fibre* $\mathbb{C}$, *the complex numbers. Then*

(a) *if* ξ_0 *is a* C_1^**-bundle over* U, *then there is a* C_1^**-bundle* ξ *over* X *unique up to* $\cong$ *such that* $\xi|X\backslash U = \varepsilon|X\backslash U$ *and* $\xi|U = \xi_0$,

(b) *if* ξ *and* ζ *are* C^**-bundles over* X *such that* $\xi|X\backslash U$ *is* *homogeneous with fibre* $\mathbb{C}$, *if* $\phi_0 : \xi|U \to \zeta$ *is a* C*-*bundle map over* U , *if* ξ *has identity section* 1 *and* q *is a projection field in* $\Gamma(\zeta)$ *with* $q|U = \phi_0 \circ (1|U)$, *then there is a unique* C*-*bundle map* $\phi : \xi \to \zeta$ *over* X *such that* $\phi|U = \phi_0$ *and* $\phi \circ 1 = q$.

Proof. The theorem 6.1 makes 6.2 an obvious special case. We now use 6.2 to prove 6.3. If ξ_0 is any C_1^*-bundle over U, then there is a unique C_1^*-bundle map $\phi : \varepsilon|U \to \xi_0$ given by the identity section of ξ_0 , so that (a) follows from 6.2(a) by taking $A = \emptyset \subset U$, as necessarily $\phi : \varepsilon|U < \xi_0$. To prove (b) we simply note that in view of (a) and Corollary 3.7, if $\xi_0 = \xi|U$, then

$$\xi \cong \varepsilon \cup_\theta \xi_0 ,$$

where $\theta : \varepsilon|U < \xi_0$ is the unique C_1^*-bundle map. Without loss of generality, we set $\xi = \varepsilon \cup_\theta \xi_0$. The projection field q defines a unique C*-bundle map $\psi : \varepsilon \to \zeta$ over X such that $\psi \circ 1_\varepsilon = q$, where 1_ε denotes the identity field of ε. Since $\phi_0 \circ (1|U) = q|U$, it follows that $\phi_0 \circ \theta = \psi|U$ and hence by 6.2(b) there is $\phi : \xi \to \zeta$ a unique C*-bundle map over X such that $\phi \circ \theta = \psi$ and $\phi|U = \phi_0$, i.e. unique such that $\phi|U = \phi_0$ and $\phi \circ 1 = q$.

We caution the reader that the total space of $\xi \cup_\phi \zeta$ in 6.2 is almost never topologically the usual adjunction space of ordinary topological pasting. In fact as soon as the open set has a boundary point, if ϕ is a proper embedding on all fibres, then the two topologies disagree. The interested reader can easily observe this phenomena in case $X = \mathbb{R}$, $A = (-\infty, 0]$, with fibres of 0 and 1 dimensions.

In 6.1 it is clear that the family of embeddings (ϕ_α) plays a role of importance in dealing with $\xi = b(z)$. We shall denote this by setting $\ell(z) = (\xi, (\phi_\alpha))$.

Suppose X and Y are both completely regular spaces, $\mathcal{U} = (U_\alpha)_{\alpha \in D}$ an open cover of X and $\mathcal{V} = (V_\alpha)_{\alpha \in D}$ an open cover of Y. We write $f : (X, \mathcal{U}) \to (Y, \mathcal{V})$ to mean $f : X \to Y$ is a continuous map such that for each $\alpha \in D$, $f(U_\alpha) \subset V_\alpha$. Let $z = (\phi_{\beta\alpha})$ be a C-cocycle for (ξ_α) over $\mathcal{U}$ and $w = (\psi_{\beta\alpha})$ a C-cocycle for (ζ_α) over $\mathcal{V}$ with $\ell(z) = (\xi, (\phi_\alpha))$ and $\ell(w) = (\zeta, (\psi_\alpha))$. Further, suppose that for each $\alpha \in D$,

$\theta_\alpha : \xi_\alpha \to \zeta_\alpha$ is a $\mathcal{C}$-bundle map covering the map $f_\alpha : U_\alpha \to V_\alpha$ defined by f, such that for $\alpha \prec \beta$,

$$(*) \qquad (\theta_\beta \mid U_{\alpha,\beta}) \circ \phi_{\beta\alpha} = \psi_{\beta\alpha} \circ (\theta_\alpha \mid U_{\alpha,\beta}).$$

In this case we denote the situation by writing simply $(\theta_\alpha) : z \to w$ covers f. If for each $\alpha \in D$, $\overline{\theta}_\alpha : f_\alpha^*(\zeta_\alpha) \to \zeta_\alpha$ is the bundle map induced by f_α, then by proposition 1.3, $(\overline{\theta}_\alpha) : f^*(w) \to w$ covers f, where $f^*(w) = (f^*(\psi_{\beta\alpha}))$.

PROPOSITION 6.4. *Let* X *and* Y *be completely regular spaces* z *a* $\mathcal{C}$*-cocycle for* $(\xi_\alpha)_{\alpha\in D}$ *over* $\mathcal{U} = (U_\alpha)_{\alpha\in D}$, w *a* $\mathcal{C}$*-cocycle for* $(\zeta_\alpha)_{\alpha\in D}$ *over* $\mathcal{V} = (V_\alpha)_{\alpha\in D}$ *with* $(\theta_\alpha)_{\alpha\in D} : z \to w$ *covering* $f : (X,\mathcal{U}) \to (Y,\mathcal{V})$. *Let* $\ell(z) = (\xi,(\phi_\alpha))$ *and* $\ell(w) = (\zeta,(\psi_\alpha))$. *Then there is* $\theta : \xi \to \zeta$ *a* $\mathcal{C}$*-bundle map covering* f *unique such that for every* $\alpha \in D$, $\theta \circ \phi_\alpha = \psi_\alpha \circ \theta_\alpha$. *If* θ_α *is a strong* $\mathcal{C}$*-bundle map for each* $\alpha \in D$, *then* θ *is a strong* $\mathcal{C}$*-bundle map. In particular, we can set* $\ell(f^*(w)) = (f^*(\zeta),(f^*(\psi_\alpha)))$.

Proof. First suppose we take the case where for each $\alpha \in D$, $\xi_\alpha = f_\alpha^*(\zeta_\alpha)$ and θ_α is the bundle map induced by f_α. It follows from the uniqueness in the factorization property of proposition 1.3 that if $\alpha \prec \beta$, then $\phi_{\beta\alpha} = f^*(\psi_{\beta\alpha})$. Let $\overline{\theta} : f^*(\zeta) \to \zeta$ be the bundle map induced by f. Then, for $\alpha \prec \beta$,

$$f^*(\psi_\beta) \circ \phi_{\beta\alpha} = f^*(\psi_\alpha)\,|U_{\alpha,\beta} : \xi_{\alpha,\beta} \to f^*(\zeta)\,|U_{\alpha,\beta}$$

so by 6.1(b), there is $\theta_1 : \xi \to f^*(\zeta)$ over X a unique $\mathcal{C}$-bundle map such that for every $\alpha \in D$, $\theta_1 \circ \phi_\alpha = f^*(\psi_\alpha)$. Set $\theta_0 = \overline{\theta} \circ \theta_1 : \xi \to \zeta$. Then θ_0 covers f. By 6.1(d), $\theta_1 : \xi \cong f^*(\zeta)$, hence θ_0 is strong. In the general case, we set $f^*(w) = (f^*(\psi_{\beta\alpha}))$. Then $f^*(w)$ is a $\mathcal{C}$-cocycle for $(f^*(\zeta_\alpha))_{\alpha\in D}$ over $\mathcal{U}$. By proposition 1.3 and our proof so far, we can set $\ell(f^*(w)) = (f^*(\zeta),(f^*(\psi_\alpha)))$, so $b(f^*(w)) = f^*(\zeta)$. Applying 1.1, for each $\alpha \in D$ there is a $\mathcal{C}$-bundle map $\delta_\alpha : \xi_\alpha \to f_\alpha^*(\zeta_\alpha)$ over U_α unique such that $\overline{\theta}_\alpha \circ \delta_\alpha = \theta_\alpha$, where $\overline{\theta}_\alpha : f_\alpha^*(\zeta_\alpha) \to \zeta_\alpha$ is the bundle map induced by f_α. The uniqueness of factorization in 1.1 guarantees that $(\delta_\alpha) : z \to f^*(w)$ covering the identity of $(X,\mathcal{U})$ so that by 6.1(b) there is $\delta : \xi \to f^*(\zeta)$ unique such that $\delta \circ \phi_\alpha = f^*(\psi_\alpha) \circ \delta_\alpha$, $\alpha \in D$. Putting $\theta = \overline{\theta} \circ \delta : \xi \to \zeta$ we have for each $\alpha \in D$, $\theta \circ \phi_\alpha = \overline{\theta} \circ \delta \circ \phi_\alpha = \overline{\theta} \circ f^*(\psi_\alpha) \circ \delta_\alpha = \psi_\alpha \circ \overline{\theta}_\alpha \circ \delta_\alpha = \psi_\alpha \circ \theta_\alpha$, so that $\theta \circ \phi_\alpha = \psi_\alpha \circ \theta_\alpha$. The uniqueness of θ follows from the corresponding uniqueness of δ and the uniqueness of the factorization of proposition 1.1. The proof is now complete.

The most useful fact in 6.4 is that $\ell(f^*(w)) = (f^*(b(w)), f^*(\psi_\alpha))$ which means in particular we can write $b(f^*(w)) = f^*(b(w))$. Therefore, our pasting procedure for pasting Banach bundles is "natural" with respect to the inducing functor.

REMARK. If the index set D is finite, then all the preceeding considerations with the obvious exception of 6.3 will remain true if $\mathcal{C} = SC^*$. Thus for an SC^*-cocycle z, the bundle $b(z)$ is easily seen to be an SC^*-bundle. In particular, 6.1, 6.2, and 6.4 are easily verified in this case. Likewise for $\mathcal{C} = SC_0^*$.

If ξ is a Hilbert bundle over X, then combining [9, proposition 21] and [15, proposition 1], we obtain an SC_0^*-bundle denoted $C^*(\xi)$ such that for each $x \in X$, the fibre of $C^*(\xi)$ over x is the C^*-algebra of compact operators on $\xi(x)$. By [15, proposition 3], this construction is natural with respect to the inducing construction. In the reverse direction, if ζ is an SC_0^*-bundle over X, then [8, 10.7.11] for each $x \in X$, there is U open in X and ξ_U a Hilbert bundle over U such that $\zeta|U \cong C^*(\xi_U)$.

Let $\phi : \xi < \zeta$ be an embedding of Hilbert bundles over X. Then ϕ defines an isomorphism $\phi_0 : \xi \cong \phi(\xi) \subset \zeta$. Let η denote the C^*-subbundle of $C^*(\zeta)$ such that $\eta(x) = \{a \in C^*(\zeta)(x) : a = q_x a q_x\}$, where q_x is the orthogonal projection in $L(\zeta(x), \zeta(x))$ whose range is $\phi(\xi(x))$. Using [8, 10.7.2], it is easy to see that η is indeed an SC_0^*-subbundle of $C^*(\zeta)$ and that restriction in each fibre defines an isomorphism $r : \eta \cong C^*(\phi(\xi))$. From [15, proposition 3], $C^*(\phi_0) : C^*(\xi) \cong C^*(\phi(\xi))$, so define $C^*(\phi) = r^{-1} \circ C^*(\phi_0)$. Then $C^*(\phi) : C^*(\xi) < C^*(\zeta)$.

If $z = (\phi_{\beta\alpha})$ is an H_1-cocycle for (ξ_α) over $\mathcal{U} = (U_\alpha)$ with respect to $\prec$, then we define $C^*(z) = (C^*(\phi_{\beta\alpha}))_\succ$, so $C^*(z)$ is an SC_0^*-cocycle. Put $\ell(z) = (\xi, (\phi_\alpha))$ and set $\ell(C^*(z)) = (\zeta, (\psi_\alpha))$. Since for $\alpha \succ \beta$ we have $C^*(\phi_\alpha) \circ C^*(\phi_{\alpha\beta}) = C^*(\phi_\beta)|U_{\alpha\beta}$, by 6.1, there is $\psi : \zeta < C^*(\xi)$ unique such that for every α, $\psi \circ \psi_\alpha = C^*(\phi_\alpha)$. But if D is finite, then it is easy to see that for each $x \in X$, $\psi(\zeta(x)) = C^*(\xi(x))$ and hence $\psi : \zeta \cong C^*(\xi)$. We state this as our next proposition.

PROPOSITION 6.5. *Let* X *be a completely regular space and* $z = (\phi_{\beta\alpha})$ *an* H_1*-cocycle for* (ξ_α) *over* $\mathcal{U} = (U_\alpha)$ *with respect to* $\prec$. *Assume that* D *is a finite set. If* $\ell(z) = (\xi,(\phi_\alpha))$, *then we can set*

$$\ell(C^*(z)) = (C^*(\xi),(C^*(\phi_\alpha))).$$

7. HOMOTOPY AND C*-BUNDLES

One of the most useful facts of ordinary fibre bundle theory is that if $f_0, f_1 : X \to Y$ are homotopic maps on the paracompact Hausdorff space X and if ξ is a fibre bundle over Y, then $f_0^*(\xi) \cong f_1^*(\xi)$. If we take $Y = X \times I$ and $f_i = \mu_i : X \to X \times I$, where $\mu_i(x) = (x,i)$, $i = 0,1$, then the general result is equivalent to the special case that if ξ is a fibre bundle over $X \times I$, then $\mu_0^*(\xi) \cong \mu_1^*(\xi)$. This in turn is equivalent to the fact that $\xi \cong \mu_0^*(\xi) \times I$. In fact, the proofs in [13, Corollary 3.5 and Corollary 3.6] show that the reduction of the general case to the special case works for any category of bundles when there is an induced bundle construction.

PROPOSITION 7.1. *Let* X *be a paracompact Hausdorff space,* ξ *a* C*-*bundle over* $X \times I$, β *a* C*-*bundle over* X, *and suppose that* ξ *and* $\beta \times I$ *are locally isomorphic. Then* $\xi \cong \mu_0^*(\xi) \times I$.

<u>Proof</u>. The proof here is a straightforward modification of the proofs in [15, lemmas 1 and 2]. We merely replace the word "Banach" by the term "C*-" throughout. The details can easily be checked by the reader.

THEOREM 7.2. *Let* X *be a metric space,* ξ *an* SC_1^*-*bundle over* $X \times I$ *having locally finite order. Then* $\xi \cong \mu_0^*(\xi) \times I$ *if and only if for each* $x \in X$, $\xi|\{x\} \times I$ *is homogeneous.*

Proof. If $\xi \cong \mu_0^*(\xi) \times I$, then for each $x \in X$, $\xi|\{x\} \times I$ is certainly homogeneous. Suppose now that for each $x \in X$, $\xi|\{x\} \times I$ is homogeneous. Then by proposition 7.1, it suffices to assume that ξ has finite order. For any SC*-bundle β, let $O(\beta)$ denote its order. We shall proceed by induction on $O(\xi)$, and assume the theorem true for any SC_1^*-bundle β over a metric space $Y \times I$ when $O(\beta) < O(\xi)$. We first observe that if $O(\xi) = 1$, then the theorem is true because both ξ and $\mu_0^*(\xi) \times I$ must be locally trivial by proposition 3.6, hence they are locally isomorphic, so by proposition 7.1, $\xi \cong \mu_0^*(\xi) \times I$. Assume that $O(\xi) > 1$. Let $x_0 \in X$ be fixed. Then $\xi|\{x_0\} \times I$ is homogeneous, hence trivial by our previous remark. There is then a trivial SC_1^*-bundle ε over X such that $\xi|\{x_0\} \times I \cong (\varepsilon|\{x_0\}) \times I$. The compactness of I combined with theorem 3.5 give the existence of an open neighborhood V of x_0 in X and an identity preserving embedding $(\varepsilon|V) \times I < \xi|V \times I$. Let β denote the image of $(\varepsilon|V) \times I$ under the embedding $(\varepsilon|V) \times I < \xi|V \times I$. Then β is a homogeneous SC_1^*-subbundle of $\xi|V \times I$, hence $\beta \cong \mu_0^*(\beta) \times I$, and moreover, if β' denotes the commutant of β in $\xi|V \times I$, then by theorem 5.3 and corollary 5.4, $\beta \otimes \beta' \cong \xi|V \times I$. Let A be the set of points $x \in V$ with $\dim \xi(x,0) = \dim \xi(x_0, 0)$. Then A is closed in V by [13, prop. 1.6], and $A \times I$ is exactly the set of points of $V \times I$ over which ξ has a fibre of dimension $\dim \xi(x_0, 0)$. Now, by our choice of β, $\beta|\{x_0\} \times I = \xi|\{x_0\} \times I$, hence $\beta|A \times I = \xi|A \times I$. This means $\beta'|A \times I$ merely has a one dimensional fibre. Let $W = V\backslash A$. Then $O(\beta'|W \times I) < O(\xi|V \times I) \leq O(\xi)$ so by our inductive hypothesis, $\beta'|W \times I \cong (\mu_0^*(\beta')|W) \times I$ and therefore by corollary 6.3, $\beta' \cong \mu_0^*(\beta') \times I$. It now follows that

$$\begin{aligned} \xi|V \times I \cong \beta \otimes \beta' &\cong (\mu_0^*(\beta) \times I) \otimes (\mu_0^*(\beta') \times I) \\ &\cong \mu_0^*(\beta \otimes \beta') \times I \\ &\cong (\mu_0^*(\xi)|V) \times I \\ &\cong (\mu_0^*(\xi) \times I)|V \times I, \end{aligned}$$

so ξ and $\mu_0^*(\xi) \times I$ are locally isomorphic and by proposition 7.1 we finally conclude that $\xi \cong \mu_0^*(\xi) \times I$.

If ξ is an SC_0^*-bundle over X, then X_∞ denotes the set of points over which ξ has infinite dimensional fibre. We shall say ξ is *proper* if all the fibres of ξ are separable and $\xi|X_\infty$ is locally trivial. If X is a metric space and X_∞ is open in X with ξ proper, then the method of proof in [13, lemma 3.9] shows that $\xi|X_\infty$ is separable, and by [13, proposition 3.7], $\xi|X\backslash X_\infty$ is separable. Since X_∞ is a countable union of closed sets in this case, ξ must be separable by [13, lemma 3.8]. On the other hand, if ξ has locally finite order, then X_∞ will have to be open. Thus any proper SC_0^*-

bundle having locally finite order and metrizable base space must in fact be separable. In the converse direction, if ξ is a separable SC_0^*-bundle over X, and if X_∞ is locally paracompact and finite dimensional, then by [8, 10.8.8], ξ must be proper. If β is a Hilbert bundle over X we recall from [14] that β is proper provided $\beta|X_\infty$ is locally trivial and each fibre of β is separable. It now follows from [8, 10.7.9] that if ξ is a proper SC_0^*-bundle over X and if $\xi \cong C^*(\beta)$, then β is a proper Hilbert bundle over X. It is also clear that if $f : X \to Y$ is a continuous map and ξ is a proper SC_0^*-bundle over Y, then $f^*(\xi)$ is a proper SC_0^*-bundle over X.

THEOREM 7.3. *Let* X *be a metric space,* ξ *a proper* SC_0^**-bundle over* $X \times I$ *having locally finite order. Then* $\xi \cong \mu_0^*(\xi) \times I$ *if and only if for each* $x \in X$, $\xi|\{x\} \times I$ *is homogeneous.*

Proof. Simply employ the method of proof of [15, proposition 4], keeping in mind our previous remarks and applying [14, proposition 4] at the appropriate place instead of the weaker result from [13].

We now wish to show that this homotopy invariance of isomorphism type also holds for any C*-bundle having only finite dimensional fibres provided it has locally at most second order.

Suppose now that ξ and ζ are Banach (C*-)bundles over X and that both are locally trivial with finite dimensional fibres. Further, suppose $i : \xi < \zeta$ is a Banach (C*-)bundle embedding. We can define a bundle map $R : L(\zeta,\zeta) \to L(\xi,\zeta)$ by restriction in each fibre so that for each section ψ of $L(\zeta,\zeta)$, we have $R \circ \psi = \psi \circ i$. Since $\|R(x)\| \leq 1$, $x \in X$ and R carries continuous sections to continuous sections, R is indeed a Banach bundle map, by 1.2. The bundle map R in turn restricts to give a bundle map $R_0 : U(\zeta) \to U(\xi,\zeta)$ and in case of C*-bundles, a bundle map $R_0 : C^*U(\zeta) \to C^*U(\xi,\zeta)$.

THEOREM 7.4. *Let* ξ *and* ζ *be homogeneous* C*-*bundles over the paracompact Hausdorff space* X, *and assume* ξ *is a* C*-*subbundle of* ζ *and that* ζ *has finite dimensional fibres. Let* $i : \xi \times I \subset \zeta \times I$ *be the inclusion and suppose* $\phi : \xi \times I < \zeta \times I$ *is any section of* $C^*U(\xi,\zeta) \times I$ *with* $\mu_0^*(\phi) = \mu_0^*(i)$. *Then there is* $\psi : \zeta \times I \cong \zeta \times I$ *over* $X \times I$ *such that* $\psi \circ i = \phi$ *and* $\mu_0^*(\psi)$ *is the identity of* ζ.

Proof. Let $R_0 : C^*U(\zeta) \to C^*U(\xi,\zeta)$ be the bundle map obtained by restricting the Banach bundle map $R : L(\zeta,\zeta) \to L(\xi,\zeta)$ as described in the previous remarks. Let Y denote the total space of $C^*U(\zeta)$ and Z the total space of $C^*U(\xi,\zeta)$ so that $R_0 : Y \to Z$. Let $Z_0 = R_0(Y) \subset Z$. We claim (R_0,Y,Z_0) is a locally trivial bundle and that Z_0 is both open and closed in Z. If $z_0 \in Z$, then z_0 projects to a point $x_0 \in X$ via projection of the bundle $C^*U(\xi,\zeta)$. By 3.11, there is an open neighborhood V of x_0 in X such that the inclusion $j : \xi|V \subset \zeta|V$ is trivial. Let F and G be the fibres of ξ and ζ respectively and let (θ,λ) be the trivialization of $j : \xi|V \subset \zeta|V$. There is then a unique injective *-homomorphism $j_0 : F < G$ such that $\lambda \circ j = \varepsilon(V,j_0) \circ \theta$. Let p be the projection of $C^*U(\zeta)$ and let q be the projection of $C^*U(\xi,\zeta)$. Set $U = p^{-1}(V)$ and $W = q^{-1}(V)$ so U and W are open subsets of Y and Z respectively and $R_0(U) \subset W$. Our trivialization (θ,λ) determines for R_0 a commutative diagram

$$\begin{array}{ccc} U & \xrightarrow{\ \tilde{\lambda}\ } & V \times \mathrm{Aut}(G) \\ \Big\downarrow {\scriptstyle R_0|U} & & \Big\downarrow {\scriptstyle id_V \times \alpha} \\ W & \xrightarrow{\ h\ } & V \times C^*U(F,G) \end{array}$$

where $\alpha : \mathrm{Aut}(G) \to C^*U(F,G)$ is the map $\alpha(g) = g \circ j_0$, $g \in \mathrm{Aut}(G)$, and where the bundle isomorphisms $\tilde{\lambda}$ and h are determined by θ and λ. Since α is a fibre bundle projection onto its image, the bundle (R_0,Y,Z_0) must be locally trivial. Denote by $\pi_2 : V \times C^*U(F,G) \to C^*U(F,G)$ the second factor projection. For $w_0 \in W$, we can denote by O the orbit of $\pi_2(h(w_0))$ under the action of $\mathrm{Aut}(G)$. By 3.10, O is open in $C^*U(F,G)$, hence $h^{-1}(V \times O)$ is an open neighborhood of w_0 which is either entirely contained in or entirely disjoint from Z_0, the image of R_0. This establishes our claim that Z_0 is both open and closed in Z and that (R_0,Y,Z_0) is a locally trivial bundle. Our section ϕ of $C^*U(\xi,\zeta) \times I$ defines a unique homotopy $f : X \times I \to Z$ which at each stage is a section of $C^*U(\xi,\zeta)$. Denoting $f_t = f \circ \mu_t$, we have $f_0(X) \subset Z_0$ because $\mu_0^*(\phi) = \mu_0^*(i)$ is just the inclusion $\xi \subset \zeta$. Since I is connected and Z_0 is both open and closed in Z, it follows that $f(X \times I) \subset Z_0$ so we have a commutative diagram

$$\begin{array}{ccc} X & \xrightarrow{\ id_\zeta\ } & Y \\ {\scriptstyle \mu_0}\downarrow & & \downarrow{\scriptstyle R_0} \\ X \times I & \xrightarrow{\ f\ } & Z_0 \end{array}$$

where $id_\zeta : X \to Y$ is of course the section of $C^*U(\zeta)$ corresponding to the identity map of ζ. Since X is paracompact Hausdorff and the bundle (R_0, Y, Z_0) is locally trivial, by the ordinary covering homotopy property for locally trivial bundles with respect to paracompact Hausdorff spaces there is a covering homotopy $\overline{f} : X \times I \to Y$ with $\overline{f}\mu_0 = id_\zeta$ and $R_0 \circ \overline{f} = f$. Since R_0 is a bundle map $R_0 : C^*U(\zeta) \to C^*U(\xi,\zeta)$ covering the identity may on X and for each $t \in I$, f_t is a section of $C^*U(\xi,\zeta)$, it follows that $\overline{f}_t = \overline{f}\mu_t$ is a section of $C^*U(\zeta)$. Consequently $\overline{f}$ defines a unique section ψ of $C^*U(\zeta) \times I$ such that $(R_0 \times id_I) \circ \psi = \phi$ which means $\psi : \zeta \times I \cong \zeta \times I$ and $\psi \circ i = \phi$. Since $\overline{f}_0 = id_\zeta$ it follows that $\mu_0^*(\psi) = id_\zeta$ and the proof is complete.

THEOREM 7.5. *Let* ξ *be a* C*-*bundle over* $X \times I$ *having only finite dimensional fibres. Assume* X *is a hereditarily paracompact Hausdorff space and that* ξ *locally is at most second order. Then* $\xi \cong \mu_0^*(\xi) \times I$ *if and only if for each* $x \in X$, $\xi | \{x\} \times I$ *is homogeneous.*

Proof. Assume that for each $x \in X$, $\xi | \{x\} \times I$ is homogeneous. By 7.1 and our hypothesis that locally ξ has at most second order, we may assume without loss of generality that ξ is at most second order. Since if ξ is of first order, 7.1 and 3.5 apply immediately to complete the proof, we assume ξ is of second order. Let F and G be finite dimensional C*-algebras such that every fibre of ξ is isomorphic to either F or G. Let $A \subset X$ be the subset of points $x \in X$ for which $\xi(x,0) \cong F$. Then $\xi | A \times I \cong (\mu_0^*(\xi)|A) \times I$ and $\xi | (X \backslash A) \times I \cong (\mu_0^*(\xi)|X \backslash A) \times I$ by 7.1 and 3.5. If both $C^*U(F,G)$ and $C^*U(G,F)$ are empty, then by 3.5 the set A must be both open and closed, hence $X \backslash A$ is also open and again 7.1 applies to complete the proof. Thus, without loss of generality, we can assume $C^*U(F,G)$ is non-empty. Since F and G are not *-isomorphic, it now follows that $\dim F < \dim G$, and hence A is a closed subset of X by [13, proposition 1.6]. Thus ξ and $\mu_0^*(\xi) \times I$ are isomorphic over the open set $(X\backslash A) \times I$ and we need only show them to be locally isomorphic over points of $A \times I$. Let $x_0 \in A$. As $\{x_0\} \times I$ is contractible, it follows that $\xi|\{x_0\} \times I$ is trivial with fibre F. We can therefore, by 3.5 and the compactness of I, find an open neighborhood V of x_0 in X and a C*-bundle embedding $\varepsilon(V \times I, F) < \xi|V \times I$. Let α be the image of this embedding, so $\alpha \subset \xi|V \times I$ is a trivial C*-subbundle. Since α is trivial, there is $\phi : \mu_0^*(\alpha) \times I \cong \alpha$, and as ξ is homogeneous over $(V\backslash A) \times I$, there is

$$\theta : (\mu_0^*(\xi)|V\backslash A) \times I \cong \xi|(V\backslash A) \times I.$$

We can also assume that $\mu_0^*(\phi)$ and $\mu_0^*(\theta)$ are identity maps of $\mu_0^*(\alpha)$ and $\mu_0^*(\xi)|V\backslash A$, respectively, by the homogeneous C*-bundle analog of [13, corollary 3.5]. We then have $\mu_0^*(\alpha)|V\backslash A \subset \mu_0^*(\xi)|V\backslash A$ and $\theta^{-1} \circ \phi_0 : (\mu_0^*(\alpha)|V\backslash A) \times I < (\mu_0^*(\xi)|V\backslash A) \times I$ where $\phi_0 = \phi|(V\backslash A) \times I$. Hence, by 7.4, there is a C*-bundle automorphism θ' of $(\mu_0^*(\xi)|V\backslash A) \times I$ such that $\mu_0^*(\theta')$ is the identity of $\mu_0^*(\xi)|V\backslash A$ and such that $\theta' \circ i = \theta^{-1} \circ \phi_0$, where i is the inclusion of $(\mu_0^*(\alpha)|V\backslash A) \times I$ in $(\mu_0^*(\xi)|V\backslash A) \times I$. Thus $\theta \circ \theta' \circ i = \phi_0 = \phi|(V\backslash A) \times I$. Using 6.2, we now notice that

$$(\mu_0^*(\xi)|V) \times I = (\mu_0^*(\alpha) \times I) \cup_i ((\mu_0^*(\xi)|V\backslash A) \times I)$$

and hence there is $\psi : (\mu_0^*(\xi)|V) \times I < \xi|V \times I$ unique such that ψ on $\alpha \times I$ is given by ϕ and $\psi|(V\backslash A) \times I = \theta \circ \theta'$. It now follows that ψ is in fact an isomorphism since the fibres are finite dimensional and for each point of $V \times I$, $(\mu_0^*(\xi)|V) \times I$ and $\xi|V \times I$ have isomorphic fibres. Thus ξ and $\mu_0^*(\xi) \times I$ are locally isomorphic and hence isomorphic by 7.1. The proof is complete.

8. PROOF OF THE THEOREM

In this section we give the proof of the main theorem. The corollary to the main theorem is then an immediate consequence of the main theorem and results of the preceeding section. We shall assume throughout this section the notation established in section one for the statement of the main theorem. If $[\phi] \in [V\backslash A, \mathcal{C}U(\beta,\xi)]$, $V \in N(W,A)$, we defined $b(\phi) = (\beta|V) \cup_\phi \xi$ which represents, by 6.2, a class in $\mathcal{CB}_X(\beta_0,\xi)$. Suppose $[\phi_0] = [\phi_1]$ in $[V\backslash A, \mathcal{C}U(\beta,\xi)]$. Then as we can naturally identify $\mathcal{C}U(\beta,\xi) \times I = \mathcal{C}U(\beta \times I, \xi \times I)$, there is $\psi \in \Gamma((V\backslash A) \times I, \mathcal{C}U(\beta \times I, \xi \times I))$ with $\mu_0^*(\psi) = \phi_0$ and $\mu_1^*(\psi) = \phi_1$. Let $\zeta = ((\beta|V) \times I) \cup_\psi (\xi \times I)$. By proposition 6.4, we have $\mu_0^*(\zeta) \cong b(\phi_0)$ and $\mu_1^*(\zeta) \cong b(\phi_1)$, hence $b(\phi_0) \underset{\mathcal{C}}{\cong} b(\phi_1)$. On the other hand, if $\phi \in \Gamma(V\backslash A, \mathcal{C}U(\beta,\xi))$, if g is a $\mathcal{C}$-bundle automorphism of $\beta|V$, and if h is a $\mathcal{C}$-bundle automorphism of ξ, then it is easy to see from 6.1(d) that $b(\phi) \cong b(h\phi g)$. Thus there is a well-defined map $b_V : \mathcal{CO}_X(V) \to \mathcal{CB}_X(\beta_0,\xi)$, for each $V \in N(W,A)$. If $V_1, V_2 \in N(W,A)$ and $j : V_1 \subseteq V_2$, then from the uniqueness in 6.1, it is clear that $b_{V_1} \circ j^* = b_{V_2}$, hence the family of maps (b_V) indexed by $N(W,A)$ defines a map $b : \mathcal{CO}_X(\beta,\xi) \to \mathcal{CB}_X(\beta_0,\xi)$ for (*) in the main theorem. We must show that the map b is a bijection.

LEMMA 8.1. *The map* b *is surjective for any paracompact Hausdorff space* X.

<u>Proof</u>. Suppose $[\zeta] \in \mathcal{CB}_X(\beta_0,\xi)$. Then $\zeta|A \cong \beta_0 \cong \beta|A$ so there is an isomorphism $\theta_1 : \beta|A \cong \zeta|A$. By theorem 3.5 and proposition 3.9, there is $V \in N(W,A)$ and $\theta \in \Gamma(V, \mathcal{C}U(\beta,\zeta))$ such that $\theta|A = \theta_1$. Also there is $\psi : \xi|X\backslash A \cong \zeta|X\backslash A$. Set $\phi = \psi^{-1} \circ \theta : \beta|V\backslash A < \xi|V\backslash A$. Then $\psi \circ \phi = \theta|V\backslash A$, hence by 6.2, $b(\phi) \cong \zeta$.

For the remainder of this section, we assume that $X \times I$ is paracompact Hausdorff.

Let Y be any space, ζ a $\mathcal{C}$-bundle over Y and $\eta \subset \zeta$ a subbundle of ζ over Y so that η is also a $\mathcal{C}$-bundle over Y. We say η is a *$\mathcal{C}$-subbundle* of ζ if the inclusion $\eta \subset \zeta$ is a $\mathcal{C}$-bundle map. Suppose now that Y is paracompact Hausdorff and K is a closed subset of Y. Suppose that η is a $\mathcal{C}$-subbundle of $\zeta|K$ over K. Then there is a unique subbundle ζ_0 of ζ over Y such that

$$\zeta_0|Y\backslash K = \zeta|Y\backslash K$$

$$\zeta_0|K = \eta$$

and hence so that

$$\zeta_0(y) = \begin{cases} \eta(y), & y \in K \\ \zeta(y), & y \in Y\backslash K. \end{cases}$$

Employing [13, Prop.1.7], 3.5 and 3.8, we see easily that ζ_0 is a $\mathcal{C}$-subbundle of ζ over Y. We call ζ_0 the $\mathcal{C}$-subbundle of ζ defined by $\eta \subset \zeta|K$, in analogy with [13, Definition 5.4]. More generally, suppose $K_1,\ldots,K_n$ are all closed subsets of Y and for each $i \le n$ we are given a $\mathcal{C}$-subbundle η_i of $\zeta|K_i$. Further, suppose that if $i,j \le n$, then $\eta_i|K_i \cap K_j = \eta_j|K_i \cap K_j$. Let ζ_1 be the $\mathcal{C}$-subbundle of ζ defined by η_1, and inductively, let ζ_{i+1} be the $\mathcal{C}$-subbundle of ζ_i defined by $\eta_{i+1} \subset \zeta_i|K_{i+1}$, $1 \le i < n$. Then ζ_n is a $\mathcal{C}$-subbundle of ζ over Y,

$$\zeta_n|Y\backslash(K_1 \cup \ldots \cup K_n) = \zeta|Y\backslash(K_1 \cup \ldots \cup K_n),$$

and for each $i \le n$, $\zeta_n|K_i = \eta_i$. We call ζ_n the *$\mathcal{C}$-subbundle of* ζ *defined by* $\eta_1,\ldots,\eta_n$.

LEMMA 8.2. *Let* $[\zeta] \in \mathcal{CB}_X(\beta_0,\xi)$ *and* $\theta_i \in \Gamma(W,\mathcal{C}U(\beta,\zeta))$, $i = 0,1$. *Then there is* $V \in N(W,A)$ *and* $\theta_1' \in \Gamma(V, \mathcal{C}U(\beta,\zeta))$ *such that* $[\theta_0|V] = [\theta_1']$ *and* $\theta_1'(\beta|V) = \theta_1(\beta)|V$.

<u>Proof</u>. Let $\eta_i = \theta_i(\beta)$, $i = 0$ or 1. Choose $V \in N(W,A)$ with $\overline{V}$, the closure of V, contained in W. Let $K = (\overline{V} \times \{0,1\}) \cup A \times I$ so that K is a closed subset of $X \times I$. Let $\tilde{\zeta}$ be the $\mathcal{C}$-subbundle of $\zeta \times I$ defined by $(\eta_0|\overline{V}) \times \{0\}$ and $(\eta_1|\overline{V}) \times \{1\}$. Then $\tilde{\zeta}|K$ is homogeneous with fibre that of β, and $\tilde{\zeta}|A \times I = (\zeta|A) \times I \cong (\beta|A) \times I$. Thus by 3.5, there is U open in $\mathbf{V} \times I$ with $A \times I \subset U$ and $h' : (\beta \times I)|\overline{U} < \tilde{\zeta}|\overline{U}$, a $\mathcal{C}$-bundle embedding. Since I is compact,

there is $V_1 \in N(V,A)$ with $V_1 \times I \subset U$. Thus, without loss of generality, we may assume $U = V_1 \times I$ and $V_1 = V$. Let $h'((\beta|\overline{V}) \times I) = \overline{\beta}$. Then $\overline{\beta}$ is a homogeneous $\mathcal{C}$-subbundle of $\tilde{\zeta}|\overline{V} \times I$, hence we are forced to conclude that $\mu_0^*(\overline{\beta}) = \eta_0|\overline{V}$ and $\mu_1^*(\overline{\beta}) = \eta_1|\overline{V}$. By the obvious $\mathcal{C}$-bundle analogue of [13, Corollary 3.5], relying on the results of the previous section, we can choose $\bar{h} : (\eta_0|\overline{V}) \times I \cong \overline{\beta}$ such that $\mu_0^*(\bar{h}) = \mathrm{id}$, the identity on $\eta_0|\overline{V}$. Then $\bar{h} : (\eta_0|\overline{V}) \times I < (\zeta|\overline{V}) \times I$ so define

$$h = \bar{h} \circ ((\theta_0|V) \times \mathrm{id}_I) : (\beta|V) \times I < (\zeta|V) \times I.$$

Since μ_0^* is a covariant functor, we find $\mu_0^*(h) = \theta_0|V$. Define $\theta_1' = \mu_1^*(h)$. We then have $[\theta_0|V] = [\theta_1']$ in $[V, \mathcal{C}U(\beta,\zeta)]$, but

$$\begin{aligned} \theta_1'(\beta|V) &= \mu_1^*(h)(\beta|V) \\ &= \mu_1^*(\bar{h})(\eta_0|V) \\ &= \mu_1^*(\overline{\beta})|V = \eta_1|V. \end{aligned}$$

But by definition of η_1 we have $\eta_1|V = \theta_1(\beta)|V$, hence $\theta_1'(\beta|V) = \theta_1(\beta)|V$, and the proof is complete.

The preceeding proof is an adaptation of the proof of the corresponding result for Hilbert bundles [14, Lemma 5]. In fact, in the proof of [14, Theorem 2], this was the only place where the Hilbert bundle structure and orthogonal complementation did not play an important role. However, due to our work in preceeding sections, we now have sufficient machinery to prove the main theorem without relying on Hilbert bundle structure as in [14].

PROOF OF THE MAIN THEOREM. We must now show that b is injective to complete the proof of the main theorem.

If $V \in N(W,A)$ and $\langle\phi\rangle \in \mathcal{CO}_X(V)$, let $\langle\langle\phi\rangle\rangle$ denote the element of $\mathcal{CO}_X(\beta,\xi)$ defined by $\langle\phi\rangle$. Now suppose that $\langle\langle\phi_i\rangle\rangle \in \mathcal{CO}_X(\beta,\xi)$, $i = 0,1$ with $b\langle\langle\phi_0\rangle\rangle = b\langle\langle\phi_1\rangle\rangle$. Suppose $V_i \in N(W,A)$ with $\phi_i \in \Gamma(V_i\backslash A, \mathcal{C}U(\beta,\xi))$, $i = 0,1$. Let $V = V_0 \cap V_1$. Since $\langle\langle\phi_i|V\backslash A\rangle\rangle = \langle\langle\phi_i\rangle\rangle$, $i = 0,1$, we may assume without loss of generality that $V_0 = V_1 = V$. We have $b(\phi_0) \underset{\mathcal{C}}{\cong} b(\phi_1)$, so there is $[\zeta] \in \mathcal{CB}_{X\times I}(\beta_0 \times I, \xi \times I)$ such that $\mu_i^*(\zeta) \cong b(\phi_i)$, $i = 0,1$. By lemma 8.1 applied to $X \times I$, $\beta_0 \times I$, $\xi \times I$, we can find a $V_2 \in N(A \times I, W \times I)$ and $\psi \in \Gamma(V_2 \backslash (A \times I), \mathcal{C}U(\beta \times I, \xi \times I))$ such that $\zeta \cong b(\psi)$. Since I is compact, there is $V_3 \in N(W,A)$ with $V_3 \times I \subset V_2$ and after again making V and V_3 smaller if necessary, we may assume $V_3 = V$, so ψ is defined over $V \times I$. We therefore have $b(\phi_i) \cong \mu_i^*(b(\psi)) \cong b(\mu_i^*(\psi))$, by 6.4, and as $[\mu_0^*(\psi)] = [\mu_1^*(\psi)]$ in $[V \backslash A, \mathcal{C}U(\beta,\xi)]$, we need only show $\langle\langle\phi_i\rangle\rangle = \langle\langle\mu_i^*(\psi)\rangle\rangle$, $i = 0,1$. Each of these problems is just the problem of showing that $\langle\langle\phi_0\rangle\rangle = \langle\langle\phi_1\rangle\rangle$ if $b(\phi_0) \cong b(\phi_1)$. We may therefore assume that $b(\phi_0) \cong b(\phi_1)$. We choose $\zeta \in [b(\phi_0)] = [b(\phi_1)]$. By 6.2 and 3.5, we have $\theta_i \in \Gamma(W_i, \mathcal{C}U(\beta,\zeta))$ with $W_i \in N(W,A)$, $i = 0,1$, and $\psi_i : \xi \cong \zeta|X\backslash A$ such that $\psi_i^{-1} \circ \theta_i = \phi_i$, $i = 0,1$. Again after making W_0, W_1, and V smaller if necessary, we may assume that $W_0 = W_1 = V$. By lemma 8.2, after another such shrinking of V, we may find $\theta_1' \in \Gamma(V, \mathcal{C}U(\beta,\zeta))$ such that $[\theta_0] = [\theta_1']$ in $[V, \mathcal{C}U(\beta,\zeta)]$ and $\theta_1'(\beta|V) = \theta_1(\beta|V)$. The remainder of the proof is now even simpler than that for [14, theorem 2] because of the machinery from section 6. Since $[\theta_0] = [\theta_1']$, it follows that $[\psi_0^{-1} \circ \theta_0] = [\psi_0^{-1} \circ \theta_1']$, that is $[\phi_0] = [\psi_0^{-1} \circ \theta_1']$. So after replacing ϕ_0 by $\psi_0^{-1} \circ \theta_1'$, and θ_0 by θ_1', we may assume that $\theta_0(\beta|V) = \theta_1(\beta|V)$. Let $\theta = \theta_1^{-1} \circ \theta_0 : \beta|V \cong \beta|V$ and $\psi = \psi_0^{-1} \circ \psi_1 : \xi \cong \xi$. Then $\psi\phi_1\theta = \psi_0^{-1}\psi_1\psi_1^{-1}\theta_1\theta_1^{-1}\theta_0 = \psi_0^{-1}\theta_0 = \phi_0$, hence $\langle\phi_0\rangle = \langle\psi\phi_1\theta\rangle = \langle\phi_1\rangle$ in $\mathcal{C}\mathcal{O}(V)$ which means $\langle\langle\phi_0\rangle\rangle = \langle\langle\phi_1\rangle\rangle$. This completes the proof of the main theorem.

At this point we remark that the same method of proof would work for any category of Banach bundles provided that the analogue of 3.5 holds for the category, namely the slight "swelling" of certain embeddings. Of course the category must have appropriate properties such as being closed under forming induced bundles, and be somehow "locally" defined in the sense that a Banach bundle locally in the category is in fact in the category, and likewise for morphisms. What we are describing is technically a fibred category satisfying appropriate sheaf properties making it a *stack* over the category of spaces [20]. The interested reader can easily work out the details [12] of the generalization which mainly amounts to problems with definitions.

In particular, everything here works for the category of Hilbert bundles in view of [14, lemma 2]. The proof here for the case of Hilbert bundles is an improvement over that of [14].

In [5] a more general definition is given for a uniform field of Banach spaces. The norm is required only to be semicontinuous. This larger category of bundles admits the inducing construction, forms a stack, and admits the pasting construction of section six, but does not satisfy 3.5, even if A is only a single point. Indeed, [13, proposition 1.6] fails for some of these more general Banach bundles.

The corollary to the main theorem allows us to classify, in particular, second order SC_1^*-bundles and proper SC_0^*-bundles over X, if X is a metric space. The C*-algebras corresponding to the SC_0^*-bundles when X is locally compact and separable are the separable continuous trace C*-algebras. It is easy to see that for compact X, the C*-algebras corresponding to the SC_1^*-bundles over X are the central CCR-algebras.

9. APPLICATIONS

Let X be a metric space. In theory the corollary to our main theorem classifies all second order C*-bundles over X whose fibres are all finite dimensional. In practical terms, however, simplifying assumptions must be made in order to make the classification computable. In this section we shall consider several simplifications and their consequences. We suppose $\mathcal{C} \in \{C^*, C_1^*, SC_1^*, SC_0^*\}$, and let A be a closed subset of X. We shall first study analogues of [14, proposition 15] and its two corollaries. If $\mathcal{G}$ is a group, we shall find it convenient to denote by $\mathcal{G}_*$ the group obtained by simply reversing the order of multiplication. To simplify notation, in this section all Banach bundles are C*-bundles, and we will abbreviate $M(m)$ by m where convenient. Thus $\varepsilon^F(X) = \varepsilon(X,F)$ denotes the product or **trivial** C*-bundle with fibre F and $\varepsilon^m(X) = \varepsilon(X,m)$ is the trivial C*-bundle with fibre $M(m)$. When there is no confusion as to the base space X, these trivial bundles will simply be denoted by ε^F or ε^m, respectively. In general, if an integer m occurs where a C*-algebra should be, it stands for $M(m)$. If F and G are C*-algebras with F finite dimensional, let $\mathcal{CB}_F^G(X,A)$ denote the set of isomorphism classes of $\mathcal{C}$-bundles over X, having fibre F over points of A and fibre G over points of $X\backslash A$, and which are locally trivial over $X\backslash A$. Of course, if G is also finite dimensional, then the condition of local triviality for these bundles over $X\backslash A$ is redundant by 3.5. If $\mathcal{C}$ is SC_0^* or SC_1^*, we naturally must also assume F and G are elementary C*-algebras. The representatives of members of $\mathcal{CB}_F^G(X,A)$ are called $\mathcal{C}(F,G)$-bundles over (X,A), or simply (F,G)-bundles when there is no confusion as to $\mathcal{C}$. We say ξ is a $\mathcal{C}$-bundle over (X,A) to mean that ξ is a $\mathcal{C}(F,G)$-bundle over (X,A) for some F and G. We define an equivalence relation $\underset{\mathcal{C}}{\approx}$ on $\mathcal{CB}_F^G(X,A)$ by setting $[\zeta_0] \underset{\mathcal{C}}{\approx} [\zeta_1]$ or $\zeta_0 \underset{\mathcal{C}}{\approx} \zeta_1$ if there is $[\zeta] \in \mathcal{CB}_F^G((X,A) \times I)$ with $\mu_0^*(\zeta) \cong \zeta_0$ and $\mu_1^*(\zeta) \cong \zeta_1$, where $\mu_t : X \to X \times I$ is the map

$\mu_t(x) = (x,t)$, $t \in I$, $x \in X$. We denote by $\Pi \mathcal{CB}_F^G(X,A)$ the set of $\tilde{\tilde{C}}$-equivalence classes of members of $\mathcal{CB}_F^G(X,A)$. Let $\mathcal{CB}_F(X) = \mathcal{CB}_F^F(X,A) = \Pi\mathcal{CB}_F^F(X,A)$ be the set of isomorphism classes of homogeneous C*-bundles with fibre F over X. We denote by Aut(F) the group of *-automorphisms of F.

If G is finite dimensional, there is no doubt as to the topology assigned to Aut(G). If G is not finite dimensional, then $\mathrm{Aut}(G) \subset L(G,G)$ and we give Aut(G) the strong operator (i.e. product) topology. As the members of Aut(G) are isometric and bijective, it is routine to see that the evaluation map $\mathrm{Aut}(G) \times G \to G$ is (jointly) continuous, that Aut(G) becomes a topological group, and consequently that the usual rule of exponential correspondence holds allowing us to identify

$$\Gamma(\mathcal{C}U(\varepsilon^G(X))) = C(X;\mathrm{Aut}(G))$$

where $C(X;\mathrm{Aut}(G))$ is simply the set of continuous maps $X \to \mathrm{Aut}(G)$. Moreover, as ξ is locally trivial with fibre G, $\mathcal{C}U(\xi)$ can be defined as the "weakly" associated fibre bundle with fibre Aut(G). Thus in the notation preceeding the statement of the main theorem, we have

$$\mathcal{CH}(\xi) = [X\backslash A, \mathcal{CU}(\xi)]$$

which somewhat simplifies the formula for $\mathcal{CO}_X(\beta,\xi)$. In particular, if $\xi \cong \varepsilon^G$, then

$$\mathcal{CH}(\xi) \cong [X\backslash A, \mathrm{Aut}(G)].$$

Also, $\mathcal{C}U(\beta,\xi)$ is locally trivial with fibre $\mathcal{C}U(F,G) \subset L(F,G)$, and as F is finite dimensional, there is no doubt as to the topology on $\mathcal{C}U(F,G)$.

We adopt the notation of the discussion preceeding the statement of the main theorem. Suppose $W \in N(X,A)$ and $[\beta] \in \mathcal{CB}_F(W)$, $[\zeta] \in \mathcal{CB}_G(X\backslash A)$. We note that our definition of $\mathcal{CO}_X(\beta,\xi,V)$ makes sense even if A is not assumed closed in X and V is any subset of W, even though V may be neither open nor a neighborhood of A. As the open neighborhoods of A form a cofinal family in the set of all neighborhoods of A, $\mathcal{CO}_X(\beta,\xi)$ is also the direct limit of the $\mathcal{CO}_X(\beta,\xi,V)$ as V ranges over all neighborhoods (not necessarily open) of A in W. Hence if $\mathcal{V}$ is any cofinal family of (not necessarily open) neighborhoods of A in W, then

$$\mathcal{CO}_X(\beta,\xi) = \varinjlim_{V \in \mathcal{V}} \mathcal{CO}_X(\beta,\xi,V).$$

Now using the covering bundle homotopy theorem for locally trivial bundles one can check that if $V_1 \subset V_2 \subset W$ and the inclusion $j : V_1 \backslash A \subset V_2 \backslash A$ is a homotopy equivalence, then

$$j^* : \mathcal{CO}_X(\beta,\xi,V_2) \to \mathcal{CO}_X(\beta,\xi,V_1)$$

is a bijection. Moreover, if X is a smooth manifold and A is a submanifold with closed tubular neighborhood T having interior T^0, $T \subset W$, then $\mathcal{CO}_X(\beta,\xi,T) \cong \mathcal{CO}_X(\beta,\xi|X\backslash T^0,T)$. We note that in this case, if ∂T denotes the boundary of T, then

$$\mathcal{CO}_X(\beta,\xi|X\backslash T^0,T) = [\partial T,\mathcal{CU}(\beta,\xi)] \Big/ ([X\backslash T^0,\mathcal{CU}(\xi)]|\partial T) \times ([T,\mathcal{CU}(\beta)]|\partial T)_*.$$

We put these observations together for our next proposition and its corollaries.

PROPOSITION 9.1. *Let* $\mathcal{V}$ *be a cofinal family of neighborhoods of* A *in* W *so each neighborhood of* A *contains some member of* $\mathcal{V}$. *Assume that if* $V_1, V_2 \in \mathcal{V}$ *then the inclusion* $V_1\backslash A \subset V_2\backslash A$ *is a homotopy equivalence. Then for each* $V \in \mathcal{V}$, *the map* $\phi \to b(\phi)$ *induces a bijection*

$$\mathcal{CO}_X(\beta,\xi,V) \cong \Pi\mathcal{CB}_X(\beta|A,\xi).$$

We point out here that because of 7.5, if G is finite dimensional then $\Pi\mathcal{CB}_X(\beta|A,\xi) = \mathcal{CB}_X(\beta|A,\xi)$ and 9.1 becomes a classification theorem for second order $\mathcal{C}$-bundles over (X,A).

COROLLARY 9.2. *Let* X *be a smooth manifold and suppose* A *is a compact submanifold of* X *with compact tubular neighborhood* T. *Let* ∂T *and* T^0 *denote respectively the boundary and interior of* T. *Then*

$$\mathcal{CO}_X(\beta,\xi|X\backslash T^0,T) \cong \Pi\mathcal{CB}_X(\beta|A,\xi)$$

and in particular, if G *is finite dimensional, then*

$$\mathcal{CB}_X(\beta|A,\xi) \cong [\partial T, \mathcal{CU}(\beta,\xi)] \Big/ ([X\backslash T^0, \mathcal{CU}(\xi)]|\partial T) \times ([T,\mathcal{CU}(\beta)]|\partial T)_*$$

COROLLARY 9.3. *Let* $X = SY$ *where* Y *is a compact metric space and suppose* $A = C^+Y$ *is the upper cone in* X. *Then*

$$\Pi\mathcal{CB}_F^G(X,A) \cong [Y, \mathcal{CU}(F,G)] \Big/ \pi_0(\mathrm{Aut}(G)) \times \pi_0(\mathrm{Aut}(F))_*.$$

Proof. As A and $X\backslash A$ are contractible, and all $\mathcal{C}$-bundles over (X,A) are locally trivial over A and over $X\backslash A$, we have

$$\Pi\mathcal{CB}_F^G(X,A) = \Pi\mathcal{CB}_X(\varepsilon^F(A), \varepsilon^G(X\backslash A)).$$

Since $\mathrm{Aut}(G)$ is a topological group and $X\backslash A$ is contractible, we have

$$[X\backslash A, \mathcal{CU}(\varepsilon^G)] = [X\backslash A, \mathrm{Aut}(G)] \cong \pi_0(\mathrm{Aut}(G))$$

and likewise for $\mathrm{Aut}(F)$. The result is now clear.

Again we point out that if $\dim G < \infty$, then by 7.5, $\Pi\mathcal{CB}_F^G(X,A) = \mathcal{CB}_F^G(X,A)$, and 9.3 classifies all second order $\mathcal{C}$-bundles over (SY, C^+Y) which have all finite dimensional fibres. In order to further clarify the formula of 9.3, we note that for any Lie group G with identity component G_0 we have

$$\pi_0(G) = G/G_0.$$

In view of these results, especially 9.3, we shall assume now that G is also finite dimensional and devote the next portion of this section to the study of $\mathcal{CU}(F,G)$.

By 3.10, $C^*U(F,G)$ is a disjoint union of coset spaces of the group $\mathrm{Aut}(G)$. We can write

$$G = \prod_{j=1}^{\ell} \mathbb{C}^{q_j} \otimes M(n_j)$$

with $1 \leq n_1 < n_2 < \ldots < n_\ell$. Since any $*$-automorphism of G must carry minimal ideals to minimal ideals and G is the product of its minimal ideals, it follows that

$$\mathrm{Aut}(G) = \prod_{j=1}^{\ell} \mathrm{Aut}(\mathbb{C}^{q_j} \otimes M(n_j)).$$

Also, $\mathrm{Aut}(\mathbb{C}^q \otimes M(n))$ is homeomorphic to $S(q) \times PU(n)^q$ where $S(q)$ is the group of permutations of $\{1,2,\ldots,q\}$ and $PU(n) = U(n)/S^1$ is the projective unitary group. Since each ideal $\mathbb{C}^q \otimes M(n_j)$ is invariant under every automorphism in $\mathrm{Aut}(G)$, it follows that if $[\xi] \in C^*\mathcal{B}_G(X)$, then there are homogeneous C*-subbundles $\xi_1,\ldots,\xi_\ell$ of ξ uniquely determined so that $\xi_1 \oplus \xi_2 \oplus \ldots \oplus \xi_\ell \cong \xi$, the isomorphism being given by addition fibrewise, and where the fibre of ξ_i is $\mathbb{C}^{q_i} \otimes M(n_i)$, $1 \le i \le \ell$. We thus have a unique decomposition for any homogemeous C*-bundle with finite dimensional fibre as a direct sum of homogeneous C*-subbundles each having a fibre of the form $\mathbb{C}^q \otimes M(n)$ for some q and n. In particular, if we have $q_j = 1$, $1 \le j \le \ell$, then each homogeneous C*-bundle with fibre G is uniquely expressible as a direct sum of homogeneous SC*-bundles. It follows that to classify homogeneous C*-bundles with fibre G, in this case, the quickest approach in actual examples is to classify the principal $PU(n_j)$ bundles over X for $1 \le j \le \ell$. In the more general situation where some $q_j > 1$, the problem of classifying homogeneous C*-bundles with fibre $\mathbb{C}^q \otimes M(n)$ arises, and as a group $\mathrm{Aut}(\mathbb{C}^q \otimes M(n))$ is the wreath product of $S(q)$ and $PU(n)^q$. Thus $\mathrm{Aut}(\mathbb{C}^q \otimes M(n))$ is not even connected and there is a natural split exact sequence

$$1 \to PU(n)^q \to \mathrm{Aut}(\mathbb{C}^q \otimes M(n)) \to S(q) \to 1$$

where the homomorphism $\mathrm{Aut}(\mathbb{C}^q \otimes M(n)) \to S(q)$ carries each automorphism to the resulting permutation of the minimal ideals of $\mathbb{C}^q \otimes M(n)$. Thus $PU(n)^q \subset \mathrm{Aut}(\mathbb{C}^q \otimes M(n))$ is the identity component of $\mathrm{Aut}(\mathbb{C}^q \otimes M(n))$ and therefore the group $\pi_0(\mathrm{Aut}(\mathbb{C}^q \otimes M(n)))$ is simply $S(q)$. Using 9.3 we have $[Y, \mathrm{Aut}(\mathbb{C}^q \otimes M(n))]/[S(q) \times S(q)_*]$ for the set of isomorphism classes of homogeneous C*-bundles over $X = SY$ with fibre $\mathbb{C}^q \otimes M(n)$. Choosing a base point $y_0 \in Y$, if $\phi : Y \to \mathrm{Aut}(\mathbb{C}^q \otimes M(n))$ and $\alpha = \phi(y_0)$, then the map $y \to \alpha^{-1}\phi(y)$ defines a C*-bundle isomorphic to $b(\phi)$. It follows that if $\langle\phi\rangle \in [Y,\mathrm{Aut}(\mathbb{C}^q \otimes M(n))]/S(q) \times S(q)_*$ we can always choose the representative ϕ so that $\phi(y_0)$ is the identity element of $\mathrm{Aut}(\mathbb{C}^q \otimes M(n))$. Let $[Y,\mathrm{Aut}(\mathbb{C}^q \otimes M(n))]_0$ denote the set of base points preserving homotopy classes of maps, where the base point of $\mathrm{Aut}(\mathbb{C}^q \otimes M(n))$ is its identity element. Since $\pi_0(\mathrm{Aut}(\mathbb{C}^q \otimes M(n))) = S(q)$, the inner automorphic action of $S(q) \subset \mathrm{Aut}(\mathbb{C}^q \otimes M(n))$ on $\mathrm{Aut}(\mathbb{C}^q \otimes M(n))$ induces an action of $S(q)$ on $[Y, \mathrm{Aut}(\mathbb{C}^q \otimes M(n))]_0$. We have a natural map

$$[Y,\mathrm{Aut}(\mathbb{C}^q \otimes M(n))]_0 / S(q) \to [Y,\mathrm{Aut}(\mathbb{C}^q \otimes M(n))] / S(q) \times S(q)_*$$

which in view of our discussion, is easily seen to be a bijection. Thus, for homogeneous C*-bundles over suspensions, our classification in 9.3 reduces to the classical formula and in particular

$$C^*\mathcal{B}_{\mathbb{C}^q \otimes M(n)}(SY) \cong [Y,\mathrm{Aut}(\mathbb{C}^q \otimes M(n))]_0 / S(q) .$$

The fact that $\mathrm{Aut}(\mathbb{C}^q \otimes M(n))$ is not path connected if $q > 1$, now distinguishes itself by producing non-trivial homogeneous C*-bundles even over the circle $S^1 = SS^0$, because

$$[S^0,\mathrm{Aut}(\mathbb{C}^q \otimes M(n))]_0 = \pi_0(\mathrm{Aut}(\mathbb{C}^q \otimes M(n))) = S(q)$$

so

$$C^*\mathcal{B}_{\mathbb{C}^q \otimes M(n)}(S^1) \cong S(q) / S(q) .$$

Since the action of $S(q)$ was induced by the inner automorphic action, in the preceeding formula $S(q)/S(q)$ is the orbit set of the inner automorphic action of $S(q)$ on itself. But this orbit set is just the set of conjugacy classes of $S(q)$, hence if $p(q)$ denotes the number of partitions of q as in number theory, then

$$\text{cardinality } C^*\mathcal{B}_{\mathbb{C}^q \otimes M(n)}(S^1) = p(q).$$

In particular, the $S(q)$ action is trivial in this special case only when $q = 2$, i.e. when $S(q)$ is abelian. The main point here is that even in the simplest examples of C*-bundles, the group action in our formulas can be non-trivial, in contrast to the situation for Hilbert bundles [13, 14] where the action is very often trivial in cases of interest (e.g. [14, proposition 2]). If we assume Y is connected, then as $PU(n)^q$ is the identity component of $\mathrm{Aut}(\mathbb{C}^q \otimes M(n))$, we have

$$[Y,\mathrm{Aut}(\mathbb{C}^q \otimes n)]_0 = [Y,PU(n)]_0^q$$

and in our formula, $S(q)$ acts on $[Y,PU(n)]_0^q$ by merely permuting coordinates. Consequently each homogeneous C*-bundle ξ with fibre $\mathbb{C}^q \otimes M(n)$ over $X = SY$, where Y is a connected space, can be uniquely decomposed as a direct sum $\xi = \xi_1 \oplus \ldots \oplus \xi_q$ where each ξ_q has fibre $M(n)$. Finally we can conclude that any homogeneous C*-bundle with finite dimensional fibre over the suspension of a compact connected space, decomposes uniquely (up to order of summands) as a direct sum of homogeneous SC_1^*-bundles. In particular, the classification of $C^*\mathcal{B}_G(SY)$ for connected Y simply reduces to the classification of principal $PU(n_j)$-bundles for $1 \leq j \leq n$, i.e. the computation of $[Y,PU(n_j)]_0$, where

$$G = \prod_{j=1}^{n} \mathbb{C}^{q_j} \otimes M(n_j).$$

The reader familiar with the classical formula for classification of principal bundles over a suspension may wonder why we require compactness in 9.3. The reason is that if $\dim F < \dim G$, our result in 3.5 only gives an embedding of ε^F over some neighborhood of C^+Y, whereas we need the embedding over some "band" neighborhood of Y. Thus our result 9.3 for homogeneous C*-bundles is not as general as the classical result.

To study $C^*U(F,G)$, it is best to represent F and G as products of their minimal ideals. If we set

$$F = \prod_{i=1}^{r} I_i \,, \; G = \prod_{j=1}^{s} J_j \,,$$

then each $\phi \in \mathrm{Hom}(F,G)$ is represented uniquely by a matrix (ϕ_{ji}) where $\phi_{ji} \in \mathrm{Hom}(I_i, J_j)$. If $p_j : G \to J_j$ and $u_i : I_i \to F$ are the coordinate projection and injection respectively, then $\phi_{ji} = p_j \phi u_i$, and since each ideal of F contains some minimal ideal, it follows that ϕ is *injective* if and only if for each i, $\phi u_i \neq 0$ and hence if and only if for each i there is j with $\phi_{ji} \neq 0$, that is each column of (ϕ_{ji}) has a non-zero entry. In any case, if $i \neq j$, then $\mathrm{Im}(\phi u_i)\mathrm{Im}(\phi u_j) = 0$, so the columns of (ϕ_{ji}) are "pairwise orthogonal".

By 3.10, if we know the components of Hom(F,G), then we know the components of C*U(F,G). But in particular, our matrix representation gives $\mathrm{Hom}(F,G) = \prod_{j=1}^{s} \mathrm{Hom}(F,J_j)$, so it suffices to consider the case where G is itself simple, say G = M(n). Now again by 3.5, to determine the components of C*U(F,n), it suffices to compute the isotropy subgroups (under the U(n)-action) of a finite number of elements of C*U(F,n) which can be selected according to the multiplicities of their matrix elements. This is simplest to do and still sufficiently illuminating in the case where F is also simple. We therefore restrict our attention now to examining the homotopy structure of C*U(m,n), $1 \le m \le n$. The ordinary unitary group U(n) acts on C*U(m,n) via its inner automorphic action on M(n). Let i_m denote the identity representation of M(m) and 0_ℓ the zero representation of dimension ℓ. If ki_m denotes the direct sum of k copies of i_m and $n = km+\ell$, then for $k \ge 1$, $ki_m \oplus 0_\ell$ can be considered as an element of C*U(m,n). Let $C^*U_k(m,n)$ denote the orbit of $ki_m \oplus 0_\ell$ under the action of U(n). It follows that if [n/m] denotes the greatest integer in n/m, then $C^*U(m,n)/U(n) \cong \{1,\dots,[n/m]\}$ and C*U(m,n) is the disjoint union of compact open sets $C^*U_k(m,n)$, $1 \le k \le [n/m]$. Consequently, if X is a connected space, then

$$[X,C^*U(m,n)] = \sum_{k=1}^{[m/n]} [X,C^*U_k(m,n)]$$

where Σ denotes disjoint union of sets.

We now see from 9.3 that since Aut(m) and Aut(n) are connected,

$$C^*\mathcal{B}_m^n(SX,CX) = \sum_{k=1}^{[n/m]} [X,C^*U_k(m,n)]$$

when X is a metric continuum. Let $C^*\mathcal{B}_{m,k}^n(SX,CX)$ denote the set of isomorphism classes of C*(m,n)-bundles over (SX,CX) corresponding to elements of $[X,C^*U_k(m,n)]$. Thus for X a metric continuum, every C*(m,n)-bundle over (SX,CX) represents an element of some

$C^*\mathcal{B}^n_{m,k}(SX,CX)$, and the integer k is a unique invariant of the bundle. Moreover, for any compact metric space we must have

(1) $$SC^*_0\mathcal{B}^n_m(SX,CX) = C^*\mathcal{B}^n_{m,1}(SX,CX)$$

and

(2) $$SC^*_1\mathcal{B}^n_m(SX,CX) = C^*\mathcal{B}^n_{m,\frac{n}{m}}(SX,CX),$$

where the right side of (2) is interpreted as the empty set if m does not divide n. Finally, our problem is to determine, as much as possible, the homotopy structure of $C^*U_k(m,n)$, $1 \le k \le [n/m]$. Since $U(n)$ acts transitively on $C^*U_k(m,n)$ we need only find the isotropy subgroup of $ki_m \oplus 0_\ell$ to determine $C^*U_k(m,n)$. If M denotes the image of $ki_m \oplus 0_\ell$ in $M(n)$, then $U \in U(n)$ belongs to this isotropy subgroup if and only if $U \in M'$, the commutant of M. That is to say U must be of the form

$$U = (1_m \otimes V) \oplus W$$

where $V \in U(k)$ and $W \in U(n-km)$. We can therefore identify the isotropy subgroup as $[1_m \otimes U(k)] \times U(n-km) \subset U(n)$, where 1_m denotes the identity of $U(m)$. Thus finally,

$$C^*U_k(m,n) = U(n)/[1_m \otimes U(k)] \times U(n-km),$$

and the lower homotopy groups of $C^*U_k(m,n)$ can be computed from those of the unitary groups via the long homotopy sequence [28]

$$\cdots\ \pi_r((1_m \otimes U(k)) \times U(n-km)) \xrightarrow{i_*} \pi_r(U(n)) \xrightarrow{p_*} \pi_r(C^*U_k(m,n))$$

$$\xrightarrow{\partial} \pi_{r-1}((1_m \otimes U(k)) \times U(n-km)) \xrightarrow{i_*} \pi_{r-1}(U(n)) \longrightarrow \cdots$$

where $i : (1_m \otimes U(k)) \times U(n-km) \to U(n)$ is the inclusion of the isotropy subgroup and where p is the natural projection of $U(n)$ onto the orbit $C^*U_k(m,n)$ of $ki_m \oplus 0_\ell$, $\ell = n-km$. For positive integers s,t with $s < t$, if $u \in U(\mathbb{C}^s,\mathbb{C}^t)$, the space of isometries of the Hilbert space $\mathbb{C}^s$ into the Hilbert space $\mathbb{C}^t$, then $\bar{u} : U(s) \to U(t)$ can be defined by $\bar{u}(v) = uvu^*+1-uu^*$, $v \in U(S)$. Obviously $\bar{u}$ is an injective homomorphism embedding $U(s)$ homeomorphically in $U(s)$, and giving $\bar{u}_* : \pi_r(U(s)) \to \pi_r(U(t))$. If $u_0, u_1 \in U(\mathbb{C}^s, \mathbb{C}^t)$, then as $U(\mathbb{C}^s, \mathbb{C}^t) = V_s(\mathbb{C}^t)$ is a Stiefel manifold and therefore path connected, $\bar{u}_0$ and $\bar{u}_1$ are homotopic so $(\bar{u}_0)_* = (\bar{u}_1)_*$. Also, for $r < 2s$, $\bar{u}_* : \pi_r(U(s)) \cong \pi_r(U(t))$, and $\pi_r(U(s))$ is 0 or $\mathbb{Z}$ according to whether r is even or odd. Since the group operations of $\pi_r(U(t))$ are induced by those of the group $U(t)$, for any t and r, if we choose $u \in U(\mathbb{C}^k, \mathbb{C}^n)$, $v \in U(\mathbb{C}^\ell, \mathbb{C}^n)$, $\ell = n-km$, then the map $i_* : \pi_r((1_m \otimes U(k)) \times U(\ell)) \to \pi_r(U(n))$ becomes the map

$$m\bar{u}_* + \bar{v}_* : \pi_r(U(k)) \times \pi_r(U(\ell)) \to \pi_r(U(n))$$

where $(m\bar{u}_* + \bar{v}_*)(\alpha,\beta) = m\bar{u}_*(\alpha) + \bar{v}_*(\beta)$. Applying these facts to our long homotopy sequence, we find that if $r < 2k$, r odd, then

$$\pi_r(C^*U_k(m,n)) = \begin{cases} \mathbb{Z}/m\mathbb{Z}, & km = n \\ 0, & km < n \text{ and } r < 2(n-km). \end{cases}$$

For even r with $r < 2k$ and with $r < 2(n-km)$ in case $km < n$, we see that $\pi_r(C^*U_k(m,n)) \cong \operatorname{Ker} i_*$. If $n = km$, then i_* is simply multiplication by m mapping $\mathbb{Z} \to \mathbb{Z}$ which is injective so $\pi_r(C^*U_k(m,n)) = 0$, whereas if $km < n$, then $\operatorname{Ker} i_*$ is the subgroup of $\mathbb{Z} \times \mathbb{Z}$ generated by $(1,-m)$ so $\pi_r(C^*U_k(m,n)) \cong \mathbb{Z}$ in this case. We can summarize the results as follows: for $1 \le r < 2k$,

$$\text{(3)} \qquad \pi_r(C^*U_k(m,n)) = \begin{cases} 0, & r \text{ odd and } r < 2(n-mk) \text{ or } r \text{ even and } n = mk \\ \mathbb{Z}/m\mathbb{Z}, & r \text{ odd and } mk = n \\ \mathbb{Z}, & r \text{ even and } r < 2(n-mk). \end{cases}$$

Evidently the SC_1^*-bundles are somewhat simpler than general SC^*-bundles, since $SC_1^*U(m,n) = C^*U_{n/m}(m,n)$. However, since for $1 \le r < 2k$, $k = n/m$,

$$\text{(4)} \qquad \pi_r(SC_1^*U(m,n)) = \begin{cases} 0, & r \text{ even} \\ \mathbb{Z}/m\mathbb{Z}, & r \text{ odd}, \end{cases}$$

even the structure of SC_1^*-bundles can be exceedingly complicated. In particular, we note the absence of any "stable" range which contrasts unfavorably with the situation for SC_0^*-bundles, the latter being the simplest of all. The functor $C^*(-)$, from Hilbert bundles to SC_0^*-

bundles, applied over a one point space gives a projection $p : V_m(\mathbb{C}^n) \to SC_0^*U(m,n)$ which is obviously the orbit space projection of the circle action on the Stiefel manifold $V_m(\mathbb{C}^n)$ given by scalar multiplication. Thus

$$SC_0^*U(m,n) = V_m(\mathbb{C}^n)/S^1.$$

The bundle $(p,V_m(\mathbb{C}^n), SC_0^*U(m,n))$ is a principal circle bundle, since the circle action on $V_m(\mathbb{C}^n)$ is free (we view $V_m(\mathbb{C}^n)$ simply as the space of linear isometries $\mathbb{C}^m \to \mathbb{C}^n$ here). The long homotopy sequence for this fibre bundle when $m < n$ gives $p_* : \pi_r(V_m(\mathbb{C}^n)) \cong \pi_r(SC_0^*U(m,n))$, $r \neq 2$ and $\pi_2(SC_0^*U(m,n)) \cong \mathbb{Z}$. In particular we have for $r \leq 2(n-m) > 0$,

$$\pi_r(SC_0^*U(m,n)) = \begin{cases} \mathbb{Z}\ , & r = 2 \\ 0\ , & r \neq 2. \end{cases} \tag{5}$$

The formula (5) in comparison with (1), (3), and (4) shows that the SC_0^*-bundles in some sense are indeed structurally simpler than SC_1^*-bundles and SC*-bundles.

A further simplification results when we take $m = 1$. Evaluation at $1 \in \mathbb{C}$ defines a homeomorphism of $SC_0^*(1,n)$ onto $G_1(\mathbb{C}^n)$, the space of rank one projections in $M(n)$. Thus $SC_0^*(1,n)$ is a complex projective space. In particular, $SC_0^*(1,2)$ is simply the 2-sphere.

We can very easily produce a simple representation for the C*-algebra $\Gamma(b(\phi))$, where $\phi : X \to C^*U(F,G)$ is a continuous map and $b(\phi)$ is the C*-bundle over SX resulting from 9.3. Define C^+X to be the upper cone in SX and C^-X to be the lower cone in SX. Let $\mathcal{A}_\phi$ be the set of pairs (f,g) such that $\phi\cdot(f|X) = g|X$. Here $\phi\cdot h(x) = \phi(x)(h(x))$ for $x \in X$. Then $\mathcal{A}_\phi$ is a C*-subalgebra of the product $C(C^+X;F) \times C(C^-X;G)$ and it is easy to see $\Gamma(b(\phi)) \cong \mathcal{A}_\phi$.

Since our theory applies to higher order C*-bundles for the cases of SC_1^* and SC_0^*, we can obtain some results for these higher order SC*-bundles if the base spaces are restricted. Let $(F_1,F_2,\ldots,F_n)$ be a family of finite dimensional C*-algebras with $F_1 \subset F_2 \subset \ldots \subset F_n$. Let $\mathcal{X} = (X;X_1,\ldots,X_n)$ where the family $(X_1,\ldots,X_n)$ partitions X so that $X_1, X_1 \cup X_2, \ldots, X_1 \cup \ldots \cup X_{n-1}$ are closed subsets of X. Put $\mathcal{F} = (F_1,F_2,\ldots,F_n)$. We say ξ is a $\mathcal{CF}$-bundle over $\mathcal{X}$ to mean ξ is a $\mathcal{C}$-bundle over X such that for each $i \leq n$, $\xi|X_i$ is homogeneous with fibre F_i. When there is no confusion as to $\mathcal{C}$, it will be dropped from the notation. We say ξ is a $\mathcal{C}$-bundle over $\mathcal{X}$ to mean ξ is a $\mathcal{C}\mathcal{F}$-bundle over $\mathcal{X}$ for some $\mathcal{F}$. The space

$$B = \bigcup_{k=1}^{n} X_k \times F_k$$

is the total space of a C*-subbundle, $\varepsilon(\mathcal{X},\mathcal{F})$, of the trivial C*-bundle $\varepsilon(X,F_n)$. We call $\varepsilon(\mathcal{X},\mathcal{F})$ the elementary $\mathcal{F}$-bundle over $\mathcal{X}$. If $\mathcal{C} = SC^*$ each F_i is $M(m_i)$ for some i, and if $\mathcal{C} = SC_0^*$, the inclusions $M(m_i) \subset M(m_{i+1})$ are all trace preserving whereas if $\mathcal{C} = SC_1^*$, then the inclusions are all unital.

By [14, proposition 17], we know that if $\dim X \le 1$, then every $SC_0^*\mathcal{F}$-bundle over $\mathcal{X}$ is isomorphic to the elementary $SC_0^*\mathcal{F}$-bundle over $\mathcal{X}$. We shall establish this result for SC_1^*-bundles.

PROPOSITION 9.4. *Let* ζ *be an* $SC_1^*(m_1,\ldots,m_n)$*-bundle over* $\mathcal{X}$ *with* $\dim X \le 1$. *Let* ξ *be a homogeneous* $C^*(k)$*-bundle over* X *and assume* k *divides* m_1. *Let* A *be closed in* X *and* $\phi \in \Gamma(A,SC_1^*U(\xi,\zeta))$. *Then there is a section of* $SC_1^*U(\xi,\zeta)$ *extending* ϕ.

Proof. Since ζ is an $SC_1^*(m_1,\ldots,m_n)$-bundle it follows by definition that m_i divides m_{i+1}, $1 \le i \le n$. Thus assuming k divides m_1 implies each fibre of $SC_1^*U(\xi,\zeta)$ is non-empty. If $x \in X_i$, then the fibre of $SC_1^*U(\xi,\zeta)$ over x is $SC_1^*U(k,m_i)$ which is a non-empty connected manifold. Thus all fibres of $SC_1^*U(\xi,\zeta)$ are path connected. Consider first the case where ζ is homogeneous. Then $SC_1^*U(\xi,\zeta)$ is a fibre bundle whose fibre we denote by Y. By 2.2, we need only consider the case where $SC_1^*U(\xi,\zeta)$ is trivial, so we must show each continuous map $A \to Y$ extends to a continuous map $X \to Y$. Since Y is a path connected manifold, this extension property is obtained by reducing the problem to the case of triangulable X using [1, (9.1) and (9.5), Chapter III]. Having now established the result when ζ is homogeneous, for the general case we proceed by induction, assuming the result holds for orders less than n. Let $K = X_1 \cup X_2 \cup \ldots \cup X_{n-1} = X \backslash X_n$. Then K is closed so there is, by the inductive hypothesis, a section ψ_1 of $SC_1^*U(\xi,\zeta)$ over K which extends $\phi|K \cap A$. As K and A are both closed, there is a section ψ_2 of $SC_1^*U(\xi,\zeta)$ over $K \cup A$ with $\psi_2|A = \phi$ and $\psi_2|K = \psi_1$. Since $K \cup A$ is closed, by normality of X and 3.5, there is an open neighborhood K_1 of $K \cup A$ and a section ψ_3 of $SC_1^*U(\xi,\zeta)$ over $\overline{K}_1$. Using the result in the homogeneous case now gives a section ψ_4 of $SC_1^*U(\xi,\zeta)$ over X_n which extends $\psi_3|\overline{K}_1 \cap X_n$. The sections $\psi_3|K_1$ and ψ_4 are defined on open sets and agree on $K_1 \cap X_n$ and thus together define a section of $SC_1^*U(\xi,\zeta)$ which extends ϕ.

THEOREM 9.5. *If* ξ *is an* $SC_1^*(m_1,\ldots,m_n)$*-bundle over* $\ast$ *with* $\dim X \le 1$, *then* ξ *is isomorphic to the elementary* $SC_1^*(m_1,\ldots,m_n)$*-bundle over* $\ast$.

Proof. Proceeding inductively, we assume the result for orders less than n. By 9.3 with $A = \emptyset$, $SC_1^*U(\varepsilon^{m_1},\xi)$ has a section thus defining an SC_1^*-embedding of ε^{m_1} into ξ whose image we denote by β. By 5.4, we have a decomposition $\xi \cong \beta \otimes \beta'$ where β' is the commutant of β in ξ. Then β' is an $SC_1^*(1,m_2/m_1,\ldots,m_n/m_1)$-bundle over $\ast$ and by the inductive hypothesis, $\beta'|X\backslash X_1$ is elementary. By 6.3, any isomorphism of $\beta'|X\backslash X_1$ onto the elementary $SC_1^*(m_2/m_1,\ldots,m_n/m_1)$-bundle over $(X\backslash X_1;X_2,\ldots,X_n)$ extends uniquely to an isomorphism of β' onto the elementary $SC_1^*(1,m_2/m_1,\ldots,m_n/m_1)$-bundle over $\ast$. Since β and β' are elementary, it follows that $\xi \cong \beta \otimes \beta'$ is elementary, and the proof is complete.

By our calculation showing that $SC_1^*(m,n)$ is not simply connected and 9.3, we see there are non-elementary second order SC_1^*-bundles over S^2. Thus theorem 9.5 would become false if the hypothesis on the dimension of X were allowed any increase without some type of topological restriction. Theorem 9.5 partially answers a question posed in [15] which is the C*-bundle analog of the question posed by Weil [21, page 84], which was mostly solved in [14, e.g. proposition 11].

The next example is of a non-elementary second order SC*-bundle over the unit interval, showing 9.4 does not hold for general SC*-bundles. Let $\phi_0 \in C^*U_1(2,4)$ and $\phi_1 \in C^*U_2(2,4)$, and set $F_0 = \phi_0(M(2))$, $F_1 = \phi_1(M(2))$. Let ξ be the subbundle of the trivial bundle $\varepsilon(I,M(4))$ whose total space is

$$\{0\} \times F_0 \cup (0,1) \times M(4) \cup \{1\} \times F_1 .$$

Then ξ cannot be elementary because any isomorphism of ξ with an elementary $SC^*(2,4)$-bundle over $(I;\{0,1\},(0,1))$ would produce a path in $C^*U(2,4)$ connecting ϕ_0 and ϕ_1 , contradicting our choice that they lie in different components of $C^*U(2,4)$. At first this might seem purely due to the fact that $\{0,1\}$ is not connected, however we can adjoin another interval at the endpoints forming the circle $S^1 = S\{0,1\}$, the suspension of $\{0,1\}$. By 9.2, we have

$$SC^*\mathcal{B}_2^4(S^1,I) = [\{0,1\},C^*U(2,4)] \cong \{0,1\}^2$$

and we see that our bundle ξ was just the restriction of a non-elementary bundle ζ in $SC^*\mathcal{B}_2^4(S^1,I)$. Of course our construction still depends on the fact that the boundary of I in S^1 is not connected, but for ζ the homogeneous parts are each over an interval. Of the four isomorphism classes of $SC^*(2,4)$-bundles over (S^1,I), clearly two are elementary and two are not. If ζ and ζ' are the two non-elementary bundles, then obviously a reflection $f : S^1 \to S^1$ which fixes the poles and interchanges $\{0,1\}$ will have $f^*(\zeta) \cong \zeta'$ and thus $\Gamma(\zeta) \cong \Gamma(\zeta')$. Thus, the four isomorphism classes of $SC^*(2,4)$-bundles over (S^1,I) give rise to precisely three isomorphism classes of C*-algebras.

10. STABLE ISOMORPHISMS

In the recent work of Brown, Douglas, and Fillmore [4, lemma 3.15] on the Ext functor, the notion of stable isomorphism for C*-algebras arises naturally, as $\mathrm{Ext}(\mathcal{A})$ depends only on the stable isomorphism class of the C*-algebra $\mathcal{A}$. Let $\mathcal{K}$ denote the C*-algebra of compact operators on the separable infinite dimensional Hilbert space H. We say C*-algebras $\mathcal{A}$ and $\mathcal{B}$ are *stably isomorphic* [2, definition 5.4], if $\mathcal{A} \otimes \mathcal{K} \cong \mathcal{B} \otimes \mathcal{K}$ and we call $\mathcal{A}$ *stable* if $\mathcal{A} \cong \mathcal{A} \otimes \mathcal{K}$. Since $\mathcal{K} \otimes \mathcal{K} \cong \mathcal{K}$, we can *stabilize* $\mathcal{A}$ by forming $\mathcal{A} \otimes \mathcal{K}$. Thus if $\mathcal{A}$ and $\mathcal{B}$ are separable C*-algebras which are stably isomorphic, then $\mathrm{Ext}(\mathcal{A}) \cong \mathrm{Ext}(\mathcal{B})$.

The recent work of Effros and Choi [16] together with recent results of Voiculescu indicate that $\mathrm{Ext}(\mathcal{A})$ is a group for $\mathcal{A}$ any separable nuclear C*-algebra. It thus seems $\mathrm{Ext}(\mathcal{A})$ will be an object of interest for C*-algebras more general than commutative C*-algebras, and we are therefore motivated to classify stable isomorphism classes of C*-algebras.

We can consider the analogous notion of stable isomorphism for C*-bundles. Consider $\mathcal{K}$ as the trivial C*-bundle over the single point 0 with fibre $\mathcal{K}$. If ξ and ζ are C*-bundles over X which are approximately locally trivial, then we say ξ and ζ are *stably isomorphic* if $\xi \check{\otimes} \mathcal{K}$ and $\zeta \check{\otimes} \mathcal{K}$ are isomorphic over $X \times 0$. Let $\varepsilon = \varepsilon(X;H)$ denote the trivial Hilbert bundle, so $C^*(\varepsilon) = \varepsilon(X;\mathcal{K})$ is a trivial SC_0^*-bundle. If $\pi_1 : X \times 0 \to X$, $\pi_2 : X \times 0 \to 0$ are the coordinate projections, then $C^*(\varepsilon) \cong \pi_2^*(K)$ as π_2 is constant, and since π_1 is a homeomorphism, on identifying $X = X \times 0$, we obtain $\xi \check{\otimes} K = \pi_1^*(\xi) \otimes C^*(\varepsilon) = \xi \otimes C^*(\varepsilon)$. Thus ξ and ζ are stably isomorphic if and only if $\xi \otimes C^*(\varepsilon) \cong \zeta \otimes C^*(\varepsilon)$. On the other hand, since $\Gamma(\mathcal{K}) = \mathcal{K}$, by 4.7 we see that if ξ and ζ are stably isomorphic C*-bundles over compact Hausdorff spaces, then $\Gamma_b(\xi)$ and $\Gamma_b(\zeta)$ are stably isomorphic C*-algebras. In particular if X is locally compact, putting zero fibre over the point at infinity shows $\Gamma_0(\xi)$ and $\Gamma_0(\zeta)$ are stably isomorphic. We can likewise say ξ is *stable* if $\xi \check{\otimes} K \cong \xi$ or equivalently, $\xi \otimes C^*(\varepsilon) \cong \xi$. Since $C^*(\varepsilon) \otimes C^*(\varepsilon) \cong C^*(\varepsilon)$ as $\varepsilon \otimes \varepsilon \cong \varepsilon$, we can *stabilize* ξ by forming $\xi \otimes C^*(\varepsilon)$. Assuming X to be a paracompact Hausdorff space, if ξ is an SC_0^*-bundle over X, then $\delta(\xi) \in \check{H}^3(X)$ is the obstruction to realizing ξ as $C^*(\beta)$ for some Hilbert bundle β over X. The invariant δ is easily seen to classify the stable isomorphism classes of separable SC_0^*-bundles over X, using properties of δ found in [7]. However, we can apply our result 9.1 to give a classification of these stable isomorphism classes over a finite dimensional double suspension which gives nice simple representatives of each stable isomorphism class, having only second order and all fibres finite dimensional.

We first examine the invariant δ more closely and give a slightly new way of defining it which is of interest. If $e \in K$ is a rank one projection, then Ke is a Hilbert space with the operator norm and left multiplication of K on Ke gives an isomorphism $K \cong C^*(Ke)$. If e_1 and e_2 are both rank-one projections in K and $u \in K$ is a partial isometry with $u^*u = e_1$, $uu^* = e_2$, then the map $a \to au^*$ is an isometry $\tilde{u} : Ke_1 \cong Ke_2$, so $C^*(\tilde{u}) : C^*(Ke_1) \cong C^*(Ke_2)$. Notice that if $w \in K$ with $w^*w = e_1$, $ww^* = e_2$ then since e_1 is rank one, there is a unique scalar $\alpha \in S^1$ with $w = \alpha u$. If $v \in K$ with $v^*v = e_2$, $vv^* = e_3$, then $\tilde{v}\tilde{u} = \widetilde{vu}$. If ξ is an SC_0^*-bundle over X having all non-zero fibres and $e \in \Gamma(U,\xi)$ is a rank one projection field, then ξe is the Banach subbundle of $\xi|U$ whose fibre over $x \in U$ is $\xi(x)e(x)$. Thus ξe is a Hilbert bundle and $\xi|U \cong C^*(\xi e)$, via the "action" of $\xi|U$ on ξe given by left multiplication. By a *fundamental system for the* SC_0^**-bundle* ξ over X we mean a triple $((U_\alpha), (e_\alpha), (u_{\alpha\beta}))$ where (U_α) is an indexed open cover of X, for each α, $e_\alpha \in \Gamma(U_\alpha,\xi)$ is a rank one projection field, and for each pair of indices α,β, $U_{\alpha\beta} = U_\alpha \cap U_\beta$, $u_{\alpha\beta} \in \Gamma(U_{\alpha\beta},\xi)$ is a partial isometry field with $u^*_{\alpha\beta}u_{\alpha\beta} = e_\beta|U_{\alpha\beta}$ and $u_{\alpha\beta}u^*_{\alpha\beta} = e_\alpha|U_{\alpha\beta}$. For each triple of indices, let $U_{\alpha\beta\gamma} = U_\alpha \cap U_\beta \cap U_\gamma$. Then $u_{\gamma\beta}u_{\beta\alpha}$ and $u_{\gamma\alpha}|U_{\alpha\beta\gamma}$ are both partial isometry fields "from" $e_\alpha|U_{\alpha\beta\gamma}$ "to" $e_\gamma|U_{\alpha\beta\gamma}$, and hence there is $z_{\gamma\beta\alpha} : U_{\alpha\beta\gamma} \to S^1$ such that $u_{\gamma\beta}u_{\beta\alpha} = z_{\gamma\beta\alpha}u_{\gamma\alpha}$. It is routine to check that $(z_{\gamma\beta\alpha})$ represents a cocycle whose cohomology class lives in $H^2(X,S)$ where S is the sheaf of germs of continuous maps to S^1. Letting R denote the sheaf of germs of continuous real valued maps, we have an exact sequence

$$0 \longrightarrow \mathbb{Z} \longrightarrow \mathcal{R} \xrightarrow{\exp 2\pi i} \mathcal{S} \longrightarrow 1$$

of sheaves, whose long exact cohomology sequence gives an isomorphism $\check{H}^2(X,\mathcal{S}) \cong \check{H}^3(X)$ via the coboundary operator, as $H^p(X,\mathcal{R}) = 0$, $p > 0$. Thus $(z_{\gamma\beta\alpha})$ defines an element $\delta(\xi) \in \check{H}^3(X)$ which is the invariant defined in [9], because $\tilde{u}_{\gamma\beta}\tilde{u}_{\beta\gamma} = z_{\gamma\beta\alpha}\tilde{u}_{\gamma\alpha}$. In fact, since $C^*(\tilde{u}_{\gamma\alpha}) = C^*(\tilde{u}_{\gamma\beta}) \circ C^*(\tilde{u}_{\beta\alpha})$, the cocycle $c = (C^*(\tilde{u}_{\alpha\beta}))$ has $b(c) \cong \xi$ by 6.1, and if $(z_{\gamma\beta\alpha})$ cobounds, the $(u_{\alpha\beta})$ can be chosen so that $h = (\tilde{u}_{\alpha\beta})$ is a cocycle and therefore by 6.5, $C^*(b(h)) \cong C^*(c) \cong \xi$. The existence of the fundamental system for ξ is simply a consequence of the fact that by definition of an SC_0^*-bundle, there are sufficiently many local rank one projection fields for their domains to cover X. The partial isometry fields of the fundamental system are then produced locally using 3.3 and a cover refining argument analogous to that in [8, 10.7.11]. On the other hand, since the "tensor product" of fundamental systems for ξ and ζ respectively over X gives a fundamental system for $\xi \otimes \zeta$, we see easily that $\delta(\xi \otimes \zeta) = \delta(\xi) + \delta(\zeta)$ as in [7]. Likewise, if ξ_* denotes the SC_0^*-bundle obtained by conjugating the scalar multiplication of ξ, clearly $\delta(\xi_*) = -\delta(\xi)$.

Suppose $\xi|U \cong C^*(\beta)$, β a separable Hilbert bundle over U. Then $\xi \otimes C^*(\varepsilon)|U \cong C^*(\beta \otimes (\varepsilon|U))$ and by [9] we have $\beta \otimes (\varepsilon|U) \cong \varepsilon|U$. Consequently $\xi \otimes C^*(\varepsilon)|U \cong C^*(\varepsilon)|U$, so every separable SC_0^*-bundle over X is *stably locally trivial*. By [8, 10.8.4], it follows that for separable SC_0^*-bundles ξ,ζ over the paracompact Hausdorff space X, we have $\xi \check{\otimes} \mathcal{K} \cong \zeta \check{\otimes} \mathcal{K}$ if and only if $\delta(\xi) = \delta(\zeta)$. And every element of $\check{H}^3(X)$ occurs as $\delta(\xi)$ for some separable SC_0^*-bundle over X. Thus $\check{H}^3(X)$ can be viewed as the set of stable isomorphism classes of separable SC_0^*-bundles over X.

We can apply these remarks to prove easily an SC_0^*-bundle analogue of [3, theorem 2.8] which has as a corollary the special case of that result for separable continuous trace C*-algebras.

PROPOSITION 10.1. *Let* X *be a paracompact Hausdorff space,* ξ *and* ζ *separable* SC_0^**-bundles over* X *having all non-zero fibres, and* $\phi : \xi < \zeta$ *an* SC_0^**-embedding over* X*. Then* ξ *and* ζ *are stably isomorphic.*

Proof. Any fundamental system for $\phi(\xi) \subset \zeta$ serves equally well as a fundamental system for ζ, and $\xi \cong \phi(\xi)$, hence $\delta(\xi) = \delta(\phi(\xi)) = \delta(\zeta)$.

COROLLARY 10.2. *Let A be a full and hereditary C*-subalgebra of the separable continuous trace C*-algebra B. Then A and B are stably isomorphic.*

Proof. Let $X = \hat{B}$. We have then a separable SC_0^*-bundle ζ over X with $B \cong \Gamma_0(\zeta)$. X is a locally compact separable metric space because B is separable with continuous trace. By [25, proposition 3.1] the subalgebra A must also have continuous trace and so corresponds to an SC*-subbundle $\xi \subset \zeta$ over X with $A = \Gamma_0(\xi)$.

We must show the inclusion $\xi \subset \zeta$ is trace preserving on fibres. Since ζ is an SC_0^*-bundle, if $x_0 \in X$ then there is an open neighborhood U of x_0 and a Hilbert bundle β over U with $\zeta|U \cong C^*(\beta)$. Since ξ is an SC_0^*-bundle (as A is continuous trace), there is a local rank one projection field e of $\xi|U$, after making U smaller if necessary. Then $e(\beta)$ is a Hilbert subbundle of β, so again after making U smaller if necessary, we can find a unit vector field v of $e(\beta)$. Then v defines a rank one projection field e' of $\zeta|U$ with $e' \leq e$. Choose a continuous function f with support inside U such that f takes the value 1 at x_0. Then fe and fe' extend to sections $\bar{e} \in A$ and $\bar{e}' \in B$ with $\bar{e}|X\backslash U = \bar{e}'|X\backslash U = 0$ and $\bar{e}|U = fe$, $\bar{e}'|U = fe'$. We then have $0 \leq \bar{e}' \leq \bar{e} \in A$, hence $\bar{e}' \in A$ and therefore as $f(x_0) = 1$, we have $e'(x_0) \in \xi(x_0)$. Since $0 \neq e'(x_0) \leq e(x_0)$ and $e(x_0)$ is rank one in $\xi(x_0)$, it follows that $e'(x_0) = e(x_0)$ and hence the inclusion $\xi(x_0) \subset \zeta(x_0)$ is indeed trace preserving. Thus ξ is an SC_0^*-subbundle of ζ and by 10.1 we have ξ and ζ stably isomorphic. Using [13, lemma 3.2], we have C*-bundles $\bar{\xi}$ and $\bar{\zeta}$ over the one point compactification $\bar{X}$ of X, extending ξ and ζ respectively, having 0 for fibre at infinity, and which must be stably isomorphic. Thus $\Gamma_b(\bar{\xi}) = \Gamma_0(\xi)$ and $\Gamma_b(\bar{\zeta}) = \Gamma_0(\zeta)$ are stably isomorphic and the proof is complete.

We now use the results of the previous section to give a classification for stable isomorphism classes of SC_0^*-bundles over a double suspension in terms of isomorphism classes of second order $SC_0^*(m,n)$-bundles (see section 9 for this definition). Let $V_m(\mathbb{C}^n)$ be the Stiefel manifold of linear isometries $\mathbb{C}^m \to \mathbb{C}^n$. The functor $C^*(-)$

from Hilbert bundles to SC_0^*-bundles over a one point space gives a map $p : V_m(\mathbb{C}^n) \to SC_0^*U(m,n)$ which we observed in the preceeding section is the orbit projection of the natural S^1 action on $V_m(\mathbb{C}^n)$, so $(p, V_m(\mathbb{C}^n), SC_0^*(m,n))$ is a principal circle bundle, as the action is free. We assume that $m < n$. If we have $X = SY$, then there is a long exact sequence [22]

$$\ldots [SY, V_m(\mathbb{C}^n)]_0 \xrightarrow{p_*} [SY, SC_0^*(m,n)]_0 \xrightarrow{\partial} [Y, S^1]_0 \xrightarrow{i_*} [Y, V_m(\mathbb{C}^n)]_0$$

where $[X',X'']_0$ denotes the set of base point preserving homotopy classes of maps of pointed spaces (we have used $[A,A']$ for free homotopy classes of maps of spaces without base points). From [27, 7.3.2], if X' has non-degenerate base point (which is the case for any pointed polyhedron), then for any simply connected, path connected space X'' with base point, $[X',X''] = [X',X'']_0$. As $V_m(\mathbb{C}^n)$ and $SC_0^*U(m,n)$ are connected, simply connected, compact manifolds, these remarks apply. Since S^1 is a path connected topological group, $[X',S^1] = [X',S^1]_0$ for any pointed space X'.

Let $\mathcal{HB}_m(X)$ and $\mathcal{HB}_m^n(X,A)$ denote respectively the set of isomorphism classes of Hilbert m-bundles over X and of Hilbert (m,n)-bundles over (X,A) as in [14]. We have for $X = S^2Y$, $A = CSY$ a diagram

$$\begin{array}{ccccccc}
\mathcal{HB}_1(X) & \longrightarrow & \mathcal{HB}_m^n(X,A) & \xrightarrow{C^*(-)} & SC_0^*\mathcal{B}_m^n(X,A) & \xrightarrow{\delta} & \check{H}^3(X) \\
\uparrow & & \uparrow & & \uparrow & & \uparrow \\
[SY, S^1]_0 & \xrightarrow{i_*} & [SY, V_m(C^n)] & \xrightarrow{p_*} & [SY, SC_0^*U(m,n)] & \xrightarrow{\partial} & [Y, S^1]
\end{array}$$

where the vertical maps are bijections, the last being the correspondence $[Y, S^1] \cong \check{H}^1(Y) \cong \check{H}^2(SY) \cong \check{H}^3(X)$. Using 6.5 the central square commutes, and with our definition of δ, the reader can check that the right hand square commutes. As $V_m(\mathbb{C}^n)$ is simply connected, the inclusion $i : S^1 \subset V_m(\mathbb{C}^n)$ into the fibre over the base point must be null homotopic. Hence $i_* = 0$ so the boundary homomorphism $\partial : [SY, SC_0^*U(m,n)] \to [Y,S^1]$ is surjective and thus δ is surjective on $SC_0^*\mathcal{B}_m^n(X,A)$. Moreover, as $\pi_k(V_m(\mathbb{C}^n)) = 0$, $k \le 2(n-m)$, it follows that for Y a compact polyhedron, with $X = S^2Y$ and $\dim Y < 2(n-m)$ we have then $[SY, V_m(\mathbb{C}^n)] = 0$ from which we conclude ∂ and δ are bijective. This proves our next result.

PROPOSITION 10.3. *Let* $X = S^2Y$ *where* Y *is a compact polyhedron and* $\dim Y < 2(n-m)$. *Let* $A = CSY$ *be the upper cone,* $A \subset X$. *Then*

$$\delta : SC_0^*\mathcal{B}_m^n(X,A) \to \check{H}^3(X)$$

is a bijection. Hence $SC_0^*(m,n)$*-bundles over* (X,A) *are isomorphic if and only if stably isomorphic, and every separable* SC_0^**-bundle over* X *is stably isomorphic to an* $SC_0^*(m,n)$*-bundle over* (X,A).

For spheres, the only case of interest naturally is the 3-sphere.

COROLLARY 10.4. *Let* D_+^3 *be the upper hemisphere in the* 3-*sphere,* S^3. *Then for* $n > 1$

$$\delta : SC_0^*\mathcal{B}_1^n(S^3,D_+^3) \to H^3(S^3)$$

is a bijection.

Recall from the preceeding section that we have a natural homeomorphism of $SC_0^*U(1,n)$ on $G_1(\mathbb{C}^n)$, the complex projective space, given by evaluation at $1 \in \mathbb{C} = M(1)$. By 10.3, $\delta : SC_0^*\mathcal{B}_1^n(S^3,D_+^3) \to H^3(S^3)$ is bijective and by 9.2 we have $SC_0^*\mathcal{B}_1^n(S^3,D_+^3) = \pi_2(G_1(\mathbb{C}^n))$. The natural inclusion $j : G_1(\mathbb{C}^m) \subset G_1(\mathbb{C}^n)$ then induces a map $\theta : SC_0^*\mathcal{B}_1^m(S^3,D_+^3) \to SC_0^*\mathcal{B}_1^n(S^3,D_+^3)$ uniquely such that for $1 < m \leq n$

$$\begin{array}{ccc} \pi_2(G_1(\mathbb{C}^m)) & \cong & SC_0^*\mathcal{B}_1^m(S^3,D_+^3) \\ \Big\downarrow{\scriptstyle j_*} & & \Big\downarrow{\scriptstyle \theta} \\ \pi_2(G_1(\mathbb{C}^n)) & \cong & SC_0^*\mathcal{B}_1^n(S^3,D_+^3) \end{array}$$

is a commutative diagram. Obviously if ξ is an $SC_0^*(1,m)$-bundle over (S^3,D_+^3) and $[\xi] = \theta([\xi])$, then $\xi < \zeta$ so by 10.1 we have $\delta(\xi) = \delta(\zeta)$. Thus the diagram

$$\begin{array}{ccc} SC_0^*\mathcal{B}_1^m(S^3,D_+^3) & \xrightarrow{\ \delta\ } & \\ \Big\downarrow{\scriptstyle \theta} & & H^3(S^3) \\ SC_0^*\mathcal{B}_1^n(S^3,D_+^3) & \xrightarrow{\ \delta\ } & \end{array}$$

is commutative. Thus j_*, θ, and both δ's are all bijections, and in particular, $\delta : SC_0^*\mathcal{B}_1^2(S^3D_+^3) \to H^3(S^3)$ is a bijection.

The phenomena of 10.4 is totally a second order phenomena because if ξ is homogeneous over X with fibre M(n), then using the observation of [24, 1.3] one can see that $\delta(\xi) \in \text{Tor}\, \check{H}^3(X)$. As $H^3(S^3) = \mathbb{Z}$, δ vanishes on homogeneous SC_0^*-bundles over S^3 having finite dimensional fibre. Also, each homogeneous part of an $SC_0^*(m,n)$-bundle over (S^3,D_+^3) is trivial so that 10.4 must be resulting merely from the fact of having two non-isomorphic fibres.

Recently Brown [3] has made a similar observation relating to the surjectivity of δ which can be applied to separable continuous trace C*-algebras having spectrum more general than a double suspension. In terms of SC_0^*-bundles and 10.1 (recall that 10.2 is a special case of the main result of [3]), the idea is fairly simple. Suppose ζ is a separable SC_0^*-bundle over X which we may as well assume stable and therefore locally trivial. Suppose $\mathcal{U}$ is an open cover of X whose nerve $N(\mathcal{U})$ has $\dim N(\mathcal{U}) < n$. For each $U \in \mathcal{U}$ assume there is a rank one projection field $e_u \in \Gamma(U,\zeta)$. There is a unique subbundle $\xi \subset \zeta$ such that for each $x \in X$, $\xi(x)$ is the elementary C*-subalgebra $q(x)\zeta(x)q(x)$, where $q(x) = \sup\{e_u(x) : x \in U \in \mathcal{U}\}$. The condition $\dim N(\mathcal{U}) < n$ guarantees that rank $q(x) \le n$ for each x and hence $\dim \xi(x) \le n^2$, for each $x \in X$. By 10.1, ξ and ζ are stably isomorphic. In particular, we can conclude that if $\dim X < n$ or if X has an open cover $\mathcal{U}$ with $\dim N(\mathcal{U}) < n$ such that $\check{H}^3(U) = 0$ for $U \in \mathcal{U}$, then every separable SC_0^*-bundle ζ over X is stably isomorphic to a finite order SC_0^*-bundle ξ over X having fibre dimension bounded above by n^2. Thus δ is surjective on this restricted class of SC_0^*-bundles. If X is a suspension, taking $\mathcal{U} = \{X\setminus\text{north pole}, X\setminus\text{south pole}\}$ we have $\dim N(\mathcal{U}) = 1$ so each separable SC_0^*-bundle is stably isomorphic to some $SC_0^*(1,2)$-bundle over X. Checking the construction in this case shows that in general the set of points where the fibre is M(2) will not contain either pole of the suspension and hence the homogeneous part with fibre M(2) can be non-trivial. Thus our result though for more restricted base space does give stronger information. In fact using the results of section 9, it is easy to write down explicitly the section algebras in $SC_0^*\mathcal{B}_1^2(S^3,D_+^3) = \pi_2(G_1(\mathbb{C}^2))$. If we denote by A_α the continuous trace C*-algebra resulting from $\alpha \in \pi_2(G_1(\mathbb{C}^2))$, it is not unreasonable to guess that

$\mathrm{Ext}(A_\alpha) \neq \mathrm{Ext}(S^3)$, for $\alpha \neq 0$. In fact from [4,5.3] we have $\mathrm{Ext}(S^3) = \mathbb{Z}$ and from [2,6.28] we have $\mathrm{Ext}(A_\alpha) = \mathrm{Ext}^{\alpha'}(S^3) = 0$, $\alpha \neq 0$, where $\alpha' = \delta(A_\alpha)$. Thus, for $\alpha \neq 0$, the C*-algebra A_α would seem to be the simplest example of a non-commutative C*-algebra A with $\mathrm{Ext}(A)$ different from $\mathrm{Ext}(\hat{A})$, and hence where non-commutativity enters in a non-trivial fashion with respect to Ext.

BIBLIOGRAPHY

1. K. Borsuk, "Theory of Retracts", Polish Scientific, Warszawa, 1967.

2. L. Brown, Extensions and the structure of C*-algebras, preprint.

3. L. Brown, Stable isomorphism of hereditary subalgebras of C*-algebras, preprint.

4. L. Brown, R. Douglas, and P. Fillmore, Extensions of C*-algebras and K-Homology, preprint.

5. J. Dauns and K. H. Hofmann, Representation of Rings by Sections, Mem. Amer. Math. Soc. 83 (1968).

6. C. Delaroche, Extensions des C*-algèbres, Bulletin de la Société Mathématique de France, Mémoire No. 29.

7. J. Dixmier, Champs continus d'espaces hilbertiennes et de C*-algèbres (II), J. Math. Pures Appl. 42 (1963), 1-20.

8. J. Dixmier, "Les C*-algèbres et leurs représentations", Gauthier-Villars, Paris, 1969.

9. J. Dixmier and A. Douady, Champs continus d'espaces hilbertiennes, Bull. Soc. Math. France 91 (1963), 227-283.

10. A. Dold, Partitions of unity in the theory of fibrations, Ann. Math., Series 2, 78 (1963), 223-255.

11. A. Douady and L. Dal Soglio-Herault, Existence de sections pour un fibre de Banach au sens de Fell, preprint.

12. M. J. Dupré, Cohomology for classifying categories of Banach bundles. Summary, preprint.

13. M. J. Dupré, Classifying Hilbert bundles, J. Functional Analysis 15 (1974), 244-278.

14. M. J. Dupré, Classifying Hilbert bundles. II, J. Functional Analysis 22 (1976), 295-322.

15. M. J. Dupré, Hilbert bundles with infinite dimensional fibres, "Recent Advances in the Representation Theory of Rings and C*-Algebras by Continuous Sections", Mem. Amer. Math. Soc. 148 (1974), 165-176.

16. M. D. Choi and E. G.Effros, The completely positive lifting problem, preprint.

17. J. M. G. Fell, The dual spaces of C*-algebras, Trans. Amer. Math. Soc., 94 (1960), 365-403.

18. J. M. G. Fell, An extension of Mackey's method to Banach *-algebraic bundles, Mem. Amer. Math. Soc. 90 (1969).

19. J. M. G. Fell, The structure of algebras of operator fields, Acta Math. 106 (1961), 233-280.

20. J. Giraud, Cohomologie non abélienne, Springer-Verlag, Berlin,1971.

21. R. Godement, Sur la théorie des représentations unitaires, Ann. Math. Series 2, 53 (1951), 68-124.

22. P. Hilton, Homotopy Theory and Duality, Gordon and Breach, New York, 1965.

23. D. Husemoller, "Fibre Bundles", McGraw-Hill, New York, 1966.

24. F. Krauss and T. Lawson, Examples of homogeneous C*-algebras, "Recent Advances in the Representation Theory of Rings and C*-Algebras by Continuous Sections", Mem. Amer. Math. Soc. 148 (1974), 153-163.

25. G. Pederson and N. Peterson, Ideals in a C*-algebra, Math. Scand. 27 (1970), 193-204.

26. S. Sakai, C*-Algebras and W*-Algebras, Springer-Verlag, New York, 1971.

27. E. Spanier, Algebraic Topology, McGraw-Hill, New York, 1966.

28. N. Steenrod, "The Topology of Fibre Bundles", Princeton University Press, Princeton, N. J., 1951.

29. M. Takesaki and J. Tomiyama, Applications of fibre bundles to the certain class of C*-algebras, Tohoku Math. J., Series 2, 13 (1961), 498-522.

30. F. J. Thayer, Liftings in the category of C*-algebras, Ph.D. thesis, Harvard University, 1972.

Added in proof:

31. J.M.G. Fell, "Induced Representations and Banach *-Algebraic Bundles", Lecture Notes in Math., No.582, Springer-Verlag, N.Y., 1977.

32. R. Godement, "Théorie des faiseaux", Hermann, Paris, 1958.

33. A. Grothendieck, "A General Theory of Fibre Spaces with Structure Sheaf", University of Kansas, 1955.

Department of Mathematics
Tulane University
New Orleans, Louisiana
70118